International Poultry Library

POULTRY EGGS

Dedicated to the Memory of
Max Butler
Who did much to develop Egg Showing

POULTRY EGGS

MALCOLM THOMPSON

Member of Council, Poultry Club of Great Britain
from 1984 to 2008

Poultry Club: Panel A Judge

Past President: Poultry Club of Great Britain

First published 2004

Revised 2017

British Library Cataloguing-in-Publication Data
A catalogue record for this book is available
from the British Library.

ISBN 1-78926-681-8

CONTENTS

PREFACE

The idea for this book was developed as Max Butler and I travelled by train from Euston to Preston where we were to judge one of the first Easter Preston Egg Shows. We both wished to promote interest in the egg sections at our Poultry Shows, but events were to frustrate our ambitions. Max died and our project remained unfulfilled.

His influence is still apparent as the judges he trained pass on to others the high standards he demanded at the show bench. I cannot ever remember hearing a disparaging word uttered about Max, who was a humble man, and a true gentleman of the Fancy.

Some of the material is adapted from articles that Max and I wrote for early issues of *Fancy Fowl* and I am grateful to the present owners for allowing the reproduction of this and some illustrations. Other material has formed the basis of the egg lectures that I have delivered to Poultry Clubs and Societies and some is being aired for the first time.

Is is thus with great pleasure that years later I dedicate this volume to Max in appreciation of the unfailing support he gave to others in the Poultry Fancy.

I also wish to record my thanks to Dr Batty who has supported the publication of this book and for his advice over the text and illustrations.

The aim of this volume is to give a detailed description of the egg and to link its science to the showing and judging of eggs – to provide a *raison d'etre* for the *Standard*. Interspersed is some history of egg showing. I hope this treatment will make for interesting reading as well as a useful reference book.

Solomons Farm, Cornwall, 2004

ACKNOWLEDGEMENTS

In addition to the acknowledgements in the Preface, I would also like to offer my thanks to the many fanciers who assisted and gave encouragement to my efforts.

There are also individuals and organizations who supplied details or illustrations. Special mention is required for the following:

1. The Poultry Club of Great Britain
2. Poultry Breed Clubs
3. British Egg Information Service
4. Authors and publishers who had already written on some aspects of eggs. The books used for reference, some of which supplied concepts or illustrations which could be adapted, are listed at the back of the book. One title in particular was used for reference and that was *The Avian Egg* by Romanoff and Romanoff, one of the great works of scientific research and application, on all types of eggs.

All authors of technical works are indebted to others who have worked in related fields. We move forward by extending our own experiences with others, thus presenting a book which is applied to a chosen area of interest; in this case the splendid field of poultry keeping and its most useful result – eggs which help to feed the world. What would we do without eggs?

This **second edition** is a response to the increased interest in Egg Showing in recent years. The text is broadly unchanged but updates have been made in the chapters dealing with Show Rules and Show Classification. A section on the Genetics of egg colour has been added to Chapter 10. In Section 3 a chapter has been included on Turkey Eggs. Finally, a new chapter has been included describing recent scientific research on the structure and composition of the egg. I am indebted to Philippe R Wilson from the University of Bath for writing this summary. The winners of the National Shows since 1973 have been updated by W Oldcorn, to whom I also thank for his unfailing support in producing this second edition.

I apologise for any omissions in these acowledgements. Needless to state, any errors or ommissions are my own.

Malcolm Thompson
Sherborne, 2015

Personalities
&
Winners

EGG SHOWMEN
Max Butler

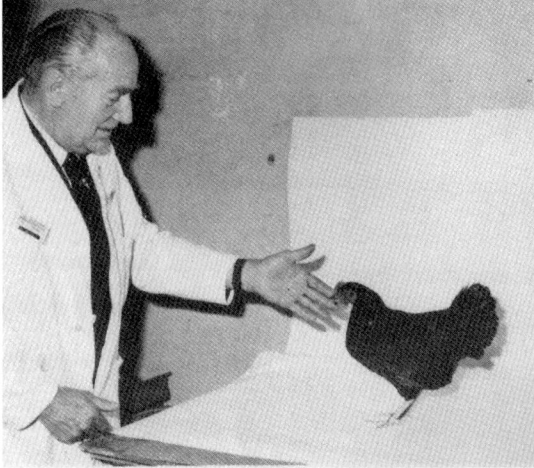

A respected Judge of poultry and eggs
A Council Member who promoted Egg Showing in the 1970's and 80's
arguing that eggs should have the same status as poultry at shows.

Max was a great supporter of white eggs. These 6 large Hamburgh
eggs were Best Eggs at Brent Show in 1973.

George Taylor

A Champion Egg Showman who holds the record for winning the most
Regional and Champion Poultry Club Egg Awards.

A successful Club Show for George at the Federation Show
with his Marans eggs.

Bill Oldcorn

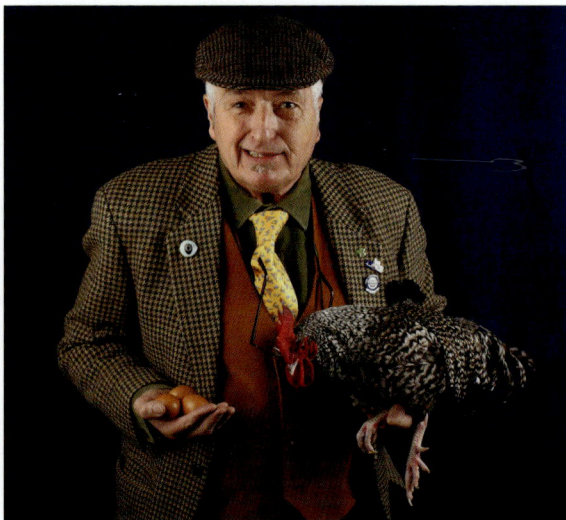

Bill with a winning Marans cockerel and Marans Eggs at the National in 2011. He has won Champion Eggs at the National more times than any other exhibitor.

A Championship win for Bill with Welsummer Eggs at the Scottish National in 2015.

WINNERS OF THE EGG SECTION
at the
NATIONAL POULTRY SHOW
1973 – 2015

1973	Miss S Colbear	6 brown/brown mottled LF
1974	F J Hatherill	3 Marans LF
1975	Miss S Colbear	3 Marans LF
1976	D T & R Ford	3 light brown LF
1977	M Butler	3 white LF
1978	A Winstanley	3 Marans LF
1979	E Boon	3 Barnevelder LF
1980	H Britcher	3 A O C LF
1981	M Butler	3 Marans bantam
1982	T Gunner	3 Marans LF
1983	**Show postponed due until February**	
1984 (Feb)	M G V Thompson	3 distinct colours bantam
1984 (Nov)	T Hill	3 white LF
1985	G Taylor	3 Marans LF
1986	D A Green	3 Marans LF
1987	G Johnson	6 Welsummer LF
1988	H Britcher	3 Marans bantam
1989	G C Taylor	3 waterfowl - duck eggs
1990	J L Blanchon	3 Marans bantam
1991	T Hill	6 A C large fowl
1992	C Pears	3 white LF
1993	P Meatyard	3 waterfowl – duck eggs
1994	G R Broadhurst	3 Barnevelder LF
1995	C Hardcastle	3 Marans bantam (See note 3)
1996	T Hill	3 light brown LF
1997	R Ashby	3 A O C bantam

1998	C & W Oldcorn	3 distinct colours LF
1999	M J Simmons	3 A O C large fowl
2000	T Hill	6 A O C bantam
2001	Rev. E Lobb	6 Welsummer LF
2002	P Hayford	6 waterfowl – goose eggs
2003	Rev D Leese	3 Marans bantam
2004	M J Simmons	3 dark brown bantam
2005	**Show cancelled due to Bird Flu**	
2006	**Show postponed to Feb 2007**	
2007 (Feb)	C & W Oldcorn	6 Marans bantam
2007 (Dec)	R G Seymour	3 distinct colours LF
2008	C & W Oldcorn	3 Marans bantam
2009	G C Taylor	3 white bantam
2010	C & W Oldcorn	3 distinct colours bantam
2011	C & W Oldcorn	6 Marans bantam
2012	T McNeight	3 Silkie LF
2013	E C Boon	6 Marans bantam
2014	C J Bennett	6 white LF
2015	E Eastham	3 distinct colours bantam

Note: The most successful exhibitors are C & W Oldcorn (5 wins); G C Taylor (3 wins); 34 years separated the two wins of E C Boon; The Marans have provided the most 'Champion' plates.

SECTION 1
THE EGG
AND ITS FORMATION

Chapter 1

WHICH CAME FIRST, THE CHICKEN OR THE EGG?

In the Beginning

The very simplest of organisms at the start of the evolutionary tree reproduced by splitting in half. This provided offspring with identical inherited materials, something that is achieved today in plants by taking cuttings or cloning. The provision of variation in offspring, so necessary in the evolutionary process, was achieved by sexual reproduction. This entailed dividing the inherited material up in 'sex cells', each having only half of the parent's materials. Random 'mating' of these sex cells produced new combinations. This inherited material is now known to be nucleic acids, mainly deoxyribose nucleic acid, D.N.A. for short. As evolution progressed the DNA was organised into chromosomes, parts of which controlled certain features, these parts are commonly called genes.

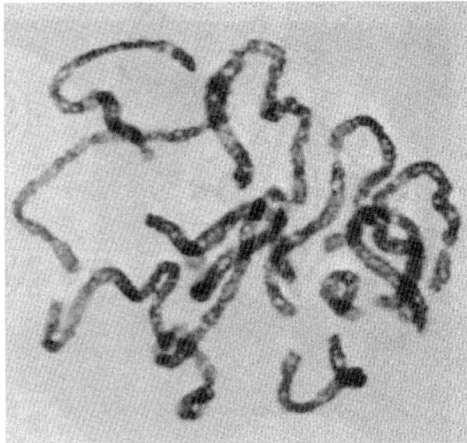

Fig 1 Chromosomes

Early Evolution

Early on in evolution the sex cells began to differ, half stayed small and mobile, half became larger and less mobile. The increase in size was due to the storing of food for the new developing organism, the embryo. Over time the sex cells, now more properly called gametes, became quite distinct, the sperm and the egg. During the millions of years that followed sperm remained small and mobile while the eggs became larger.

However, there is a limit to egg size and as organisms became more complex the egg could not store sufficient food to produce a new adult. In some situations the egg provided enough food to establish a 'half' formed but independent organism, the larva. An example is the caterpillar of the butterfly. In other situations the mother retained the egg after fertilisation and provided food for the developing embryo. Such is the case with the mammals, the most advanced having developed a placenta to enable exchange of food and waste.

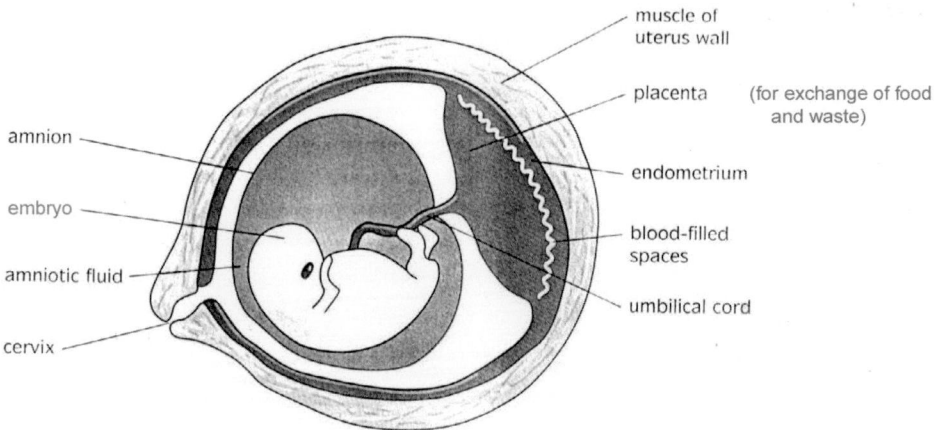

Fig 2 Mammalian Placenta

In addition to the requirement of providing food for the embryo there was also the need for safety, safety from both predators and the environment. In the water, where most early life lived, the sperm could swim to the egg, the water providing support and a stable medium in which to survive. On land eggs were subjected to extremes of temperature and drying out. The amphibians, the first large creatures to live on land, returned to the water to breed protecting their eggs in 'jelly'. The comparatively small egg developing into a larval form, the tadpole, which is also protected by the watery medium. If organisms are to reproduce on land successfully they must have solutions to the transfer of sperm, the provision of sufficient food to create an independent offspring and protection from drying up and predators.

The Land Egg

The solutions were to be found in the land egg. This was first developed by the reptiles, which includes the dinosaurs, and then perfected by the birds. This provided an enormous reserve of food in both yolk and albumen together with a bag that filled with water in which the embryo develops – it's own 'pre-natal' pond. The whole structure is enclosed in a rigid shell, which is perforated by very fine pores that enable air to reach the developing chick. Waste gases can pass out through the shell and waste solids are stored in a special sac within the egg.

Reptiles protect their eggs from desiccation and predators by hiding the egg in the ground or composting vegetation. This also serves to provide an even heat for incubation.

In birds, behaviour patterns have evolved to ensure the safety of the egg, such as nest building and brooding. Two distinct strategies have been developed – nest haters and nest lovers. The former include quail, pheasants and poultry, the latter blackbirds, tits and birds of prey. In poultry the chick when hatched is within a few hours able to run, feed

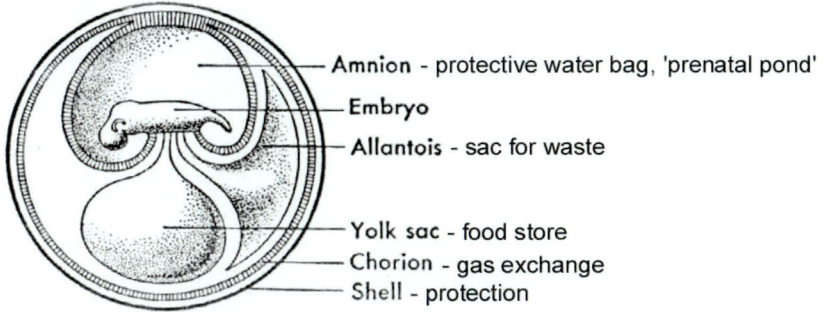

Amnion - protective water bag, 'prenatal pond'
Embryo
Allantois - sac for waste

Yolk sac - food store
Chorion - gas exchange
Shell - protection

Diagramatic land egg evolved by reptiles

Fig 3 Diagrammatic Land Egg

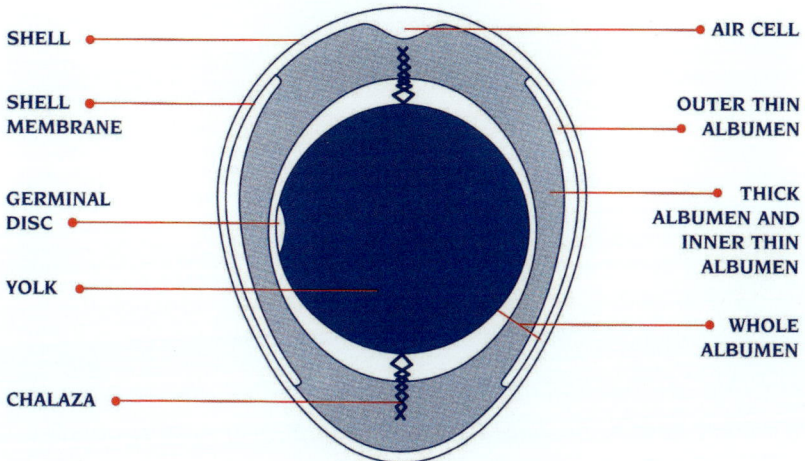

SHELL

SHELL MEMBRANE

GERMINAL DISC

YOLK

CHALAZA

AIR CELL

OUTER THIN ALBUMEN

THICK ALBUMEN AND INNER THIN ALBUMEN

WHOLE ALBUMEN

Fig 4 An Egg in cross section

and respond to its mother's call. Nest lovers by contrast are hatched blind, featherless and unable to feed. They need a prolonged period of parental care in the nest. These quite distinct early life styles demand different types of egg. To produce a chick 'up and running' requires greater food reserves and an egg shell that can survive the longer incubation.

Pictures showing early development within the egg

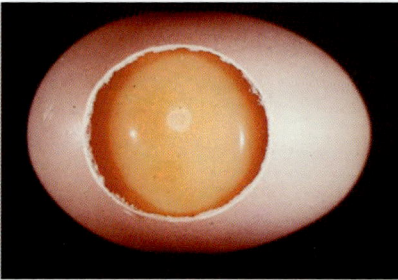

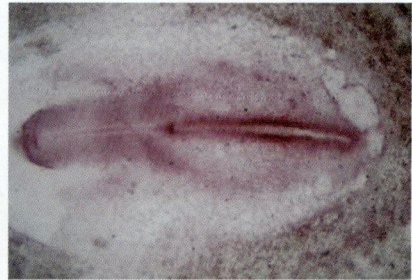

Germinal disc   **Primitive streak 18-24 hours**

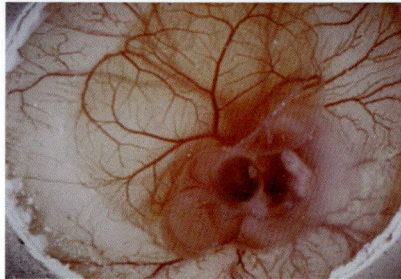

Embryo at 5-6 days showing blood vessels on yolk, developing eye in centre, to left the head, to right two white patches – limb buds surrounding heart

Hopefully this introduction has served to show how the eggs of domestic poultry have been shaped by their evolutionary past. It should also answer the question posed at the start, 'Which came first, the chicken or the egg?' Undeniably the egg!

Chapter 2

WHAT'S IN AN EGG?

Terms Used

Before exploring the formation of the egg it is only right to mention that the egg is not what many think it to be – to be biologically correct the female gamete is called the ovum (plural ova). This is the cell that carries the inherited material from the mother. In birds this cell is the yolk of the egg. The yolk of the Ostrich egg forms the largest cell in the natural world.

The Yolk

The yolk or ovum develops in a single ovary, a unique feature of birds, for most animals have two. Stimulated by increasing day length the bird produces hormones that promote the growth of the ova and the cell divisions which distribute the chromosomes. Each ovum contains half the number of chromosomes possessed by the rest of the cells of the adult bird. When fertilised by the sperm the chromosome number is reconstituted. In poultry there are 39 pairs of chromosomes in the adults. This includes a pair of sex chromosomes, XX in cocks and XY in hens. It is the ovum, therefore, that dictates the sex of the chick as half the ova contain the X chromosome and half the Y.

The chromosomes of the ovum are found on the surface of the yolk in what remains of the nucleus – the blastodisc. This disc contains cell fluid (cytoplasm) and is surrounded by granules of 'white' yolk. In the unfertilised egg this can be seen as a small chalky looking speck. The whole yolk is enclosed in the vitelline membrane. When a yolk is examined on the verge of being hard boiled, concentric layers can be recognised – the 'white' yolk and the 'yellow' yolk. The former is lighter

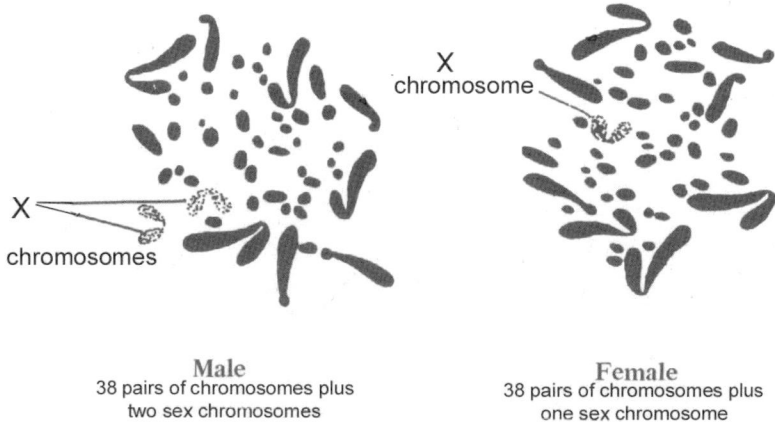

X
chromosome

X
chromosomes

Male
38 pairs of chromosomes plus
two sex chromosomes

Female
38 pairs of chromosomes plus
one sex chromosome

**Fig 5a Male and Female Chicken Chromosomes
as seen under the microscope**

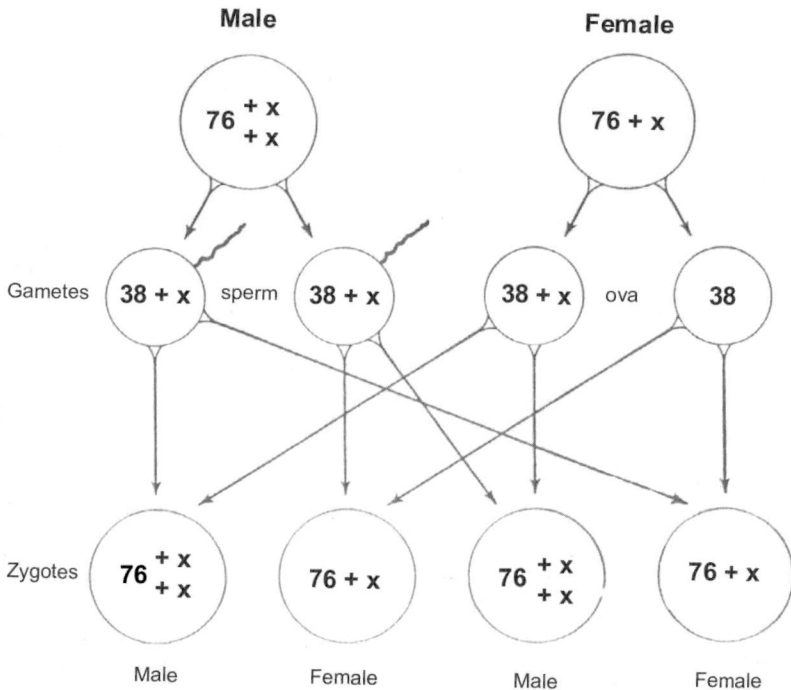

Male

Female

$76 \begin{array}{c} + x \\ + x \end{array}$

$76 + x$

Gametes $38 + x$ sperm $38 + x$ $38 + x$ ova 38

Zygotes $76 \begin{array}{c} + x \\ + x \end{array}$ $76 + x$ $76 \begin{array}{c} + x \\ + x \end{array}$ $76 + x$

Male   Female   Male   Female

**Fig 5b Diagram showing how the Chromosome number remains
constant through the generations**

in colour and forms inner rings within the yolk. It also surrounds the blastodisc and forms a 'plug' beneath it which reaches to the centre of the yolk – the latebra. White yolk contains little pigment and has a high water content, it is laid down slowly, long after feeding has ceased i.e. 1 to 5 a.m. By contrast the yellow yolk is laid down quickly during and immediately after feeding and is rich in pigments, lipids (fats) and proteins. There are usually about six broad rings of yellow yolk.

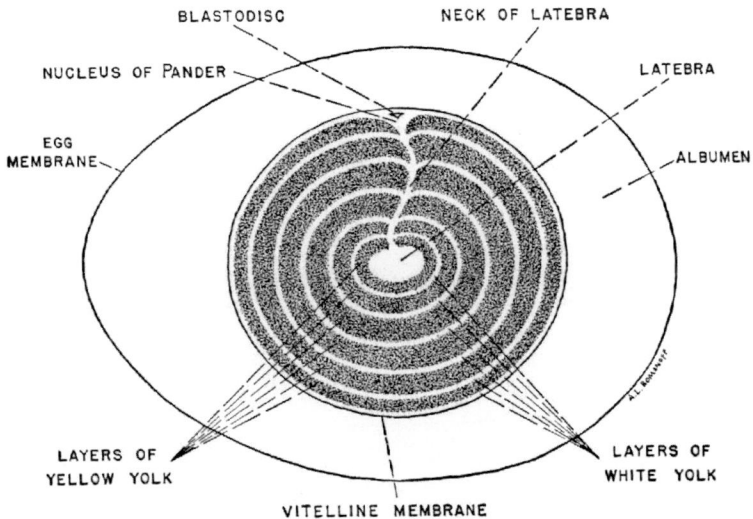

Fig 6 Ovum (yolk) showing the layers

The number of ova visible in a laying hen's ovary varies, over 2000 have been counted. The number that enlarge and develop depends upon the feeding and health of the bird. The laying record is 361 eggs in 364 days held by a black Orpington. The older the hen the slower the maturing, so fewer eggs are laid. Growth is slow for months but in the 6 to 10 days prior to being released from the ovary the yolk increases 25 fold adding up to 4mm to its diameter in a day. The materials needed

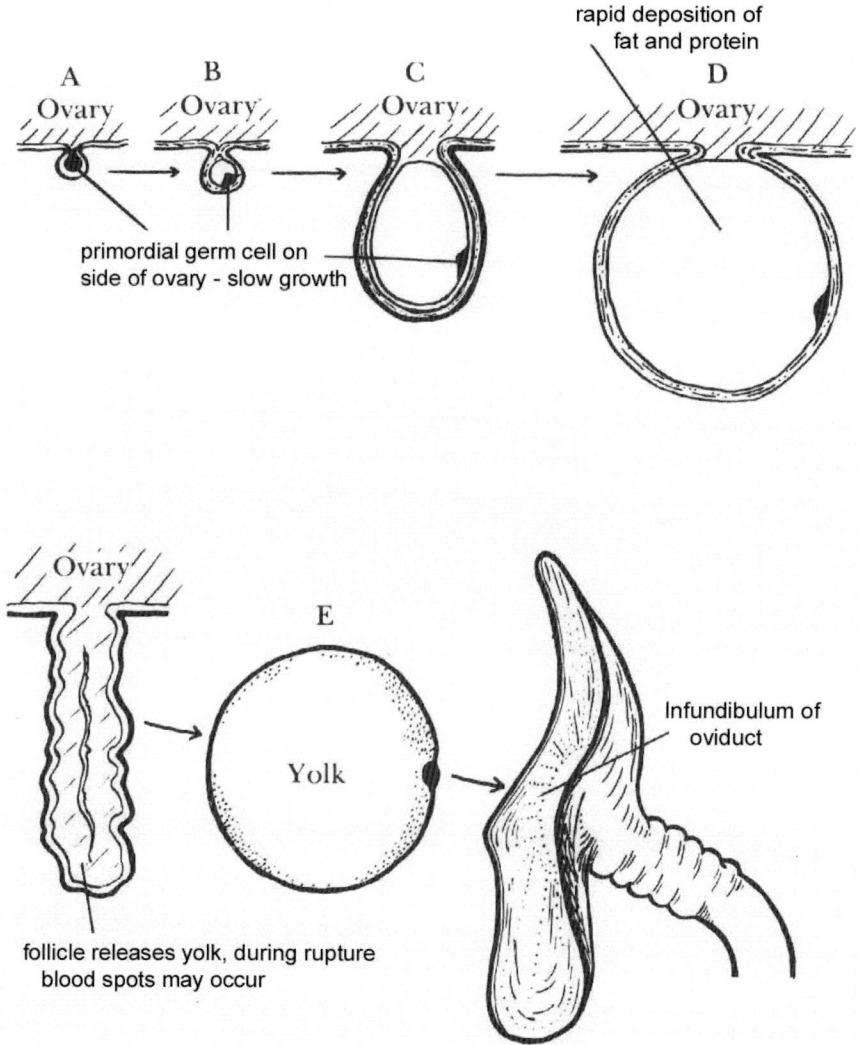

Fig 7 Development of ovum

for growth are supplied by blood vessels which run on the surface of the follicle. This is the membrane that encloses the developing ovum within the ovary. It acts to aid the exchange of materials from the blood to the yolk. In the process of ovulation this membrane ruptures and the ovum, surrounded by its elastic vitelline membrane, is engulfed by the neck of the oviduct. It is during this rupture that small blood vessels in the follicle can burst smearing the yolk with blood, a most unwelcome sight when the egg is cracked open.

Mating and Fertilisation

Mating is a regular occurrence in the poultry yard. The cock may 'tit bit' to attract the hens attention. This involves pecking at the ground while making short staccato sounds. When approached a willing hen will adopt a passive squating pose that allows a brief 'treading' and close apposition of sex organs. Sperms enter the cloaca which is the common opening for anus and urogenital system. At mating the entrance to the oviduct is everted and moved to a central position within the cloaca.

Sperm can reach the top of the oviduct within 30 minutes and they can survive there for 10-14 days, so one mating can provide for several released ova. When ova enter the neck of the oviduct one sperm will penetrate the vitelline membrane. The chromosomes in the head of the sperm will mix with those in the blastodisc of the ovum. After fertilisation cell divisions start to occur which continue as the 'egg' passes down the oviduct. The passage down the oviduct takes between 18 and 36 hours, by which time the blastoderm has developed. This can be 2-4mm across and consists of a central transparent spot in an opaque surround. Development then stops until brooding starts. However, eggs left in nest boxes can be heated by other hens when they lay their eggs. This heat can cause the growth of the blastoderm to continue. Such an enlarged blastodisc is very unsightly when the egg is used for cooking.

The Albumen

During the passage of the fertilised ovum down the oviduct first the albumen and then the outer egg membranes and shell are added. The albumen, like the yolk, consists of two distinct types, 'dense and fluid', arranged in four concentric layers.

Immediately surrounding the yolk is the chalaziferous layer. This is dense and is mixed with fine mucin fibres which at each end of the yolk are gathered and twisted into the chalazae. The chalaza at the pointed end of the egg is the longer and larger of the two and consists of two major strands twisted anti-clockwise. Whereas the chalaza at the blunt end consists of a single strand twisted clockwise. Their ends are held in the main portion of albumen and they serve to stabilise the position of the yolk within the egg and keep the blastodisc uppermost. Rotation is made possible because the yolk and inner chalaziferous layer are separated from the main dense albumen by a thin liquid layer of albumen. The main dense area of albumen makes up 60% of the total and when an egg is broken open it holds its shape due to mucin fibres and supports the yolk in its centre. As well as providing food for the developing chick it also acts as a cushion against knocks. It is attached at both ends to the inner shell membrane. This is surrounded by the outermost layer of albumen which is fluid and spreads when the egg is cracked open.

Egg Membranes

External to the albumen lies a double membrane, the inner called the egg membrane is tightly fixed to the outer, the shell membrane, everywhere except at the blunt end of the egg. Here they part to form the air sac, which provides air for the chick at hatching. Both membranes have minute pores which allow for the diffusion of gases both in and out of the egg. They consist of protein fibres such as keratin which are held in place by albuminous cement. The outer membrane is very closely attached to the shell, as it is onto its surface that the salts of calcium carbonate are deposited.

The Shell

The thickness of the shell varies with the season – it is thicker in winter – and with the eggs position in the laying cycle. The shell is thickest at the sharp end. Three layers are recognised within the shell itself – the innermost **mammillary layer**, the **spongy layer** and the **cuticle**.

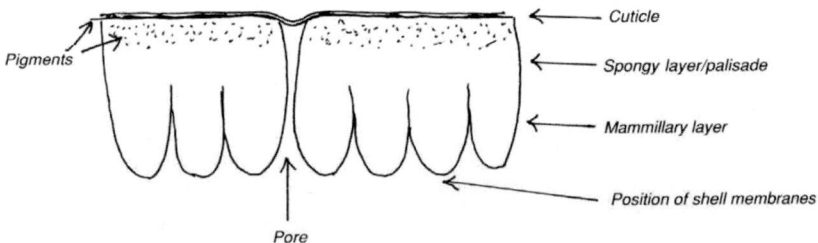

Fig 8 The Structure of the Egg Shell

The *mammillary layer* consists of oval knobs which are compressed together. As a result of their uneven shape there are many air spaces on the inner side.These connect to the pore canals that transverse the spongy layer. Both these layers consist mainly of calcium, and magnesium salts deposited on a protein matrix. Far from being soft the spongy layer is hard and compact giving rigidity and strength to the shell.

The ground colour of the egg is provided by pigments in the spongy layer. The final layer is the cuticle, a thin coating of protein which covers all the shell including the pores. It is permeable to gases so they can still enter the egg but it keeps out unwanted bacteria. Vigorous cleaning of eggs which damages this cuticle is ill advised and does nothing for the 'shelf life' of the egg. The cuticle provides the so-called 'bloom' of a hen's egg, the shine that indicates freshness. Within the cuticle lie pigments whose colour are specific to different species and whose patterns are specific to individual birds.

LITTLE TERN   COMMON SANDPIPER   RINGED PLOVER

BLACK-HEADED GULL   STONE CURLEW   LAPWING

RED GROUSE   GOLDEN PLOVER   MERLIN

Colour patterns on the shells of different birds

The foregoing account has outlined the structure of the egg. Each part is dealt with in more detail in subsequent chapters. Before this, however, it is appropriate to consider a little more fully the laying of an egg.

Chapter 3

EGG FORMATION

How It Occurs

The diagram shows the one functioning oviduct and ovary of the hen, the left one. In cases of a disease or malfunction of this ovary, the right one can develop, but it functions to produce male hormones and so cause 'sex changes' in the hen.

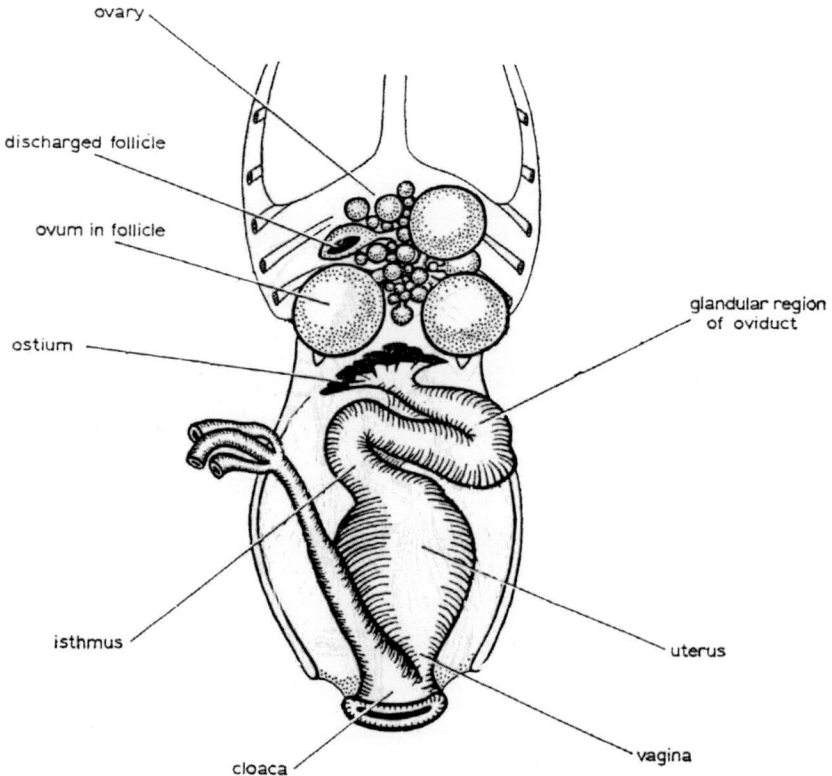

Fig 9 The Ovary and Oviduct of the Hen

Infundibulum

The cell division which produces the reduced number of chromosomes in the ovum (meiosis) is only completed after the penetration of the vitelline membrane by the sperm. The ovum by this time has usually broken free of the nurturing follicle and has been engulfed by the infundibular orifice. This part of the oviduct is suspended by tissue attachments and it moves within the body cavity so enhancing the likelihood of a released ovum gaining entrance.

The Magnum

After the thin walled entrance of the infundibulum the walls of the oviduct become thicker and the diameter narrower. In this portion, the magnum, the longest in the duct, glandular cells in the walls secrete the albumen, first the chalaziferous layer then the thick albumen. The yolk is moved down the oviduct by the muscular contractions of the wall – peristalsis. The inner wall is highly folded, which increases the surface area for secretion. These folds are spirally arranged so rotating the egg as it proceeds. This rotation serves to tighten mucin fibres in the albumen which firstly results in the formation of the chalazae. After the secretion of the thick albumen the twisting of fibres squeezes fluid from the matrix so forming the inner fluid layer. This fluid enables the yolk balanced by the chalazae to rotate within the thick albumen. The end of the magnum or albumen secreting region is marked by a ring of non-glandular rather translucent tissue.

The Isthmus

A comparatively short section follows, the isthmus. The walls are very thick, darker in colour than any other section and the inner layer is not so folded. These folds are not spirally arranged. As the egg enters this section tubular glands secrete granules of a keratin like substance which coalesce into sticky fibres which cover the albumen. These tightly applied fibres produce the inner membrane. After a pause the egg proceeds down the rest of the isthmus where the second quite distinct, shell membrane, is deposited. It is formed in a similar manner but the fibres are not quite so dense and it does have a feint pink tint due to the secretion of ooporphyrin.

The Uterus

The rather inaptly named uterus or shell gland follows. This region is capable of great expansion and encloses the egg as the shell is deposited. This involves tubular cells depositing mainly calcium salts and protein. The deposits begin as small granules on the shell membrane which become the cone shaped mammillae of the mammillary layer. The mammillae touch at their tips but small holes are left in places that connect to the larger air spaces beneath. On top of the cones a collagen like protein forms a framework on to which more calcium salts are deposited. The process is slow and it is not until 18 hours have elapsed that the 'spongy' layer is complete, having used about 5 grams of calcium salts. Pores through this layer connect to the air spaces at the surface of the mammillary layer so enabling ventilation down to the egg membranes. In addition to calcium salts pigments from the breakdown of blood products in the liver are infused into the layer to produce ground colour.

Once the shell has been completed the last section of the uterus can apply pigment in quantity to the surface as blotches or speckles. These are then covered and fixed by the application of the cuticle. The colour can serve to camouflage the egg, the general colour and pigment pattern being characteristic for each bird species. Precise patterns are provided by individual birds and related to the distribution of their secretory cells in the last part of the uterus. In poultry the pigment is secreted with the cuticle and if the egg is laid quickly after this the colour will smudge or wipe off on the nest material. This is often seen in dark brown egg layers such as Welsummers.

The Vagina

The Vagina completes the oviduct, its thick muscular walls expelling the egg to the outside. No glandular cells or secretions are present here. The egg is normally expelled pointed end first from the everted uterus, the hen standing for the final pushes. If the blunt end arrives first it is because the pointed end became caught in bottom of the uterus which has the effect of turning the egg over. The table below gives the average length of the parts of the oviduct and the time spent in these sections by the developing egg.

Oviduct part	Length mm	Formation of:	Approx time hours
Infundibulum	90	–	¼
Magnum	320	Thick albumen, chalazae	3
Isthmus	100	Thick albumen, shell membranes	1½
Shell gland	120	Thin albumen, shell, cuticle	20

Table 1 Showing length of Oviduct parts, what is formed in each part and the estimated time spent in each part

The average length of an oviduct in a laying hen is about 70 cm which can reduce to 17 cm in full moult, a quarter of the active length. The average time for an egg to travel from the ovary to the outside is about 24 hours. The acidity of the duct increases from 6.4 in the infundibulum and magnum to 5.8 in the isthmus and uterus. The higher pH aids sperm survival, while the lower pH provides the optimum conditions for enzymes controlling Calcium deposition.

Chapter 4

FACTORS AFFECTING EGG PRODUCTION

Nature or Nurture?

Egg production is determined by genetics and the environment. Genetics determines the breed and the strain. The environment includes the hens rearing regime, current feeding, housing, day length and health. The following diagram summarises the various interactions that affect the number of eggs laid by fowls.

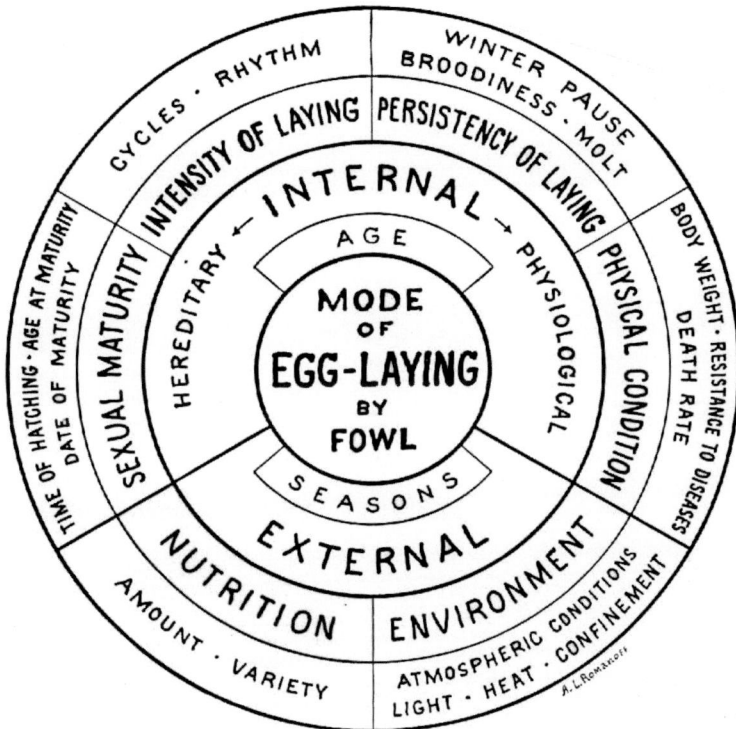

Fig 10 Factors that affect the number of eggs laid by fowls

It must be borne in mind that most of the birds behaviour is instinctive not intelligent. They are genetically programmed to respond in certain ways to their environment and situation. Their 'learned' behaviour is imprinted and conditioned using a part of the brain called the corpus striatum. The cerebral hemispheres which are the centre of mammals adaptive behaviour and of abstract thought in humans are not developed in birds. They are thus very governed by the environment in which they find themselves.

The amount that the genes affect a feature shown by an animal is expressed as a heritability percentage. The higher the percentage the greater the genetic influence.

Table 2 Heritability percentage of features shown by the chicken

Feature	Heritability
Egg size	40-50 %
Egg shape	25-50 %
Shell colour	30-90 %
Yolk colour	10-40 %
Albumen firmness	10-70 %
Blood spot frequency	10-50 %
Number of eggs laid	5-10 %

It can be seen that egg production is predominantly under environmental control. Husbandry therefore is all important.

Number of Eggs Laid

The number of eggs laid in the wild is quite constant for species, e.g. Pigeon 2 eggs, Robin 5-6 and Pheasant 10-12. This optimum number is the result of selection over time balancing the cost to the species in terms of energy input to reproduction with its other life processes and the survival rate. Generally large birds lay few eggs and small birds a lot of eggs. Pheasants and hens are an exception, possibly because their newly hatched mobile young are very susceptible to predators and other environmental dangers. In the wild, birds lay clutches of eggs. The eggs in the nest stimulate nerves on the female's brood pouches and broody behaviour is initiated. Removing eggs from the nest prevents broodiness and extends laying. Selective breeding for extended laying during domestication has eliminated broodiness in breeds such as theLeghorns and Anconas.

Frequency of Laying

Many hens show a laying rhythm e.g. two days laying, one off or five days laying, one off. This is related to the length of the laying process which if longer than 24 hours will make the hen lay later each day so prompting a blank day as eggs ready in late afternoon are retained for the following morning.

Table 3 Times of laying in a free range flock

Time	Hens Laying (per cent)
7 A.M. to 9 A.M.	17.7
9 A.M. to 11 A.M.	28.5
11 A.M. to 1 P.M.	27.3
1 P.M. to 3 P.M.	19.5
3 P.M. to 5 P.M.	7.0

Effect of hormones

The maturation of ova and their release is under the control of hormones whose secretion is influenced by the amount and intensity of light, specifically the amount of U.V

Left to nature it is the seasons which control laying. To achieve continuous laying the seasonal effects must be nullified. This laying can then be maximised by providing ideal feeding and living conditions

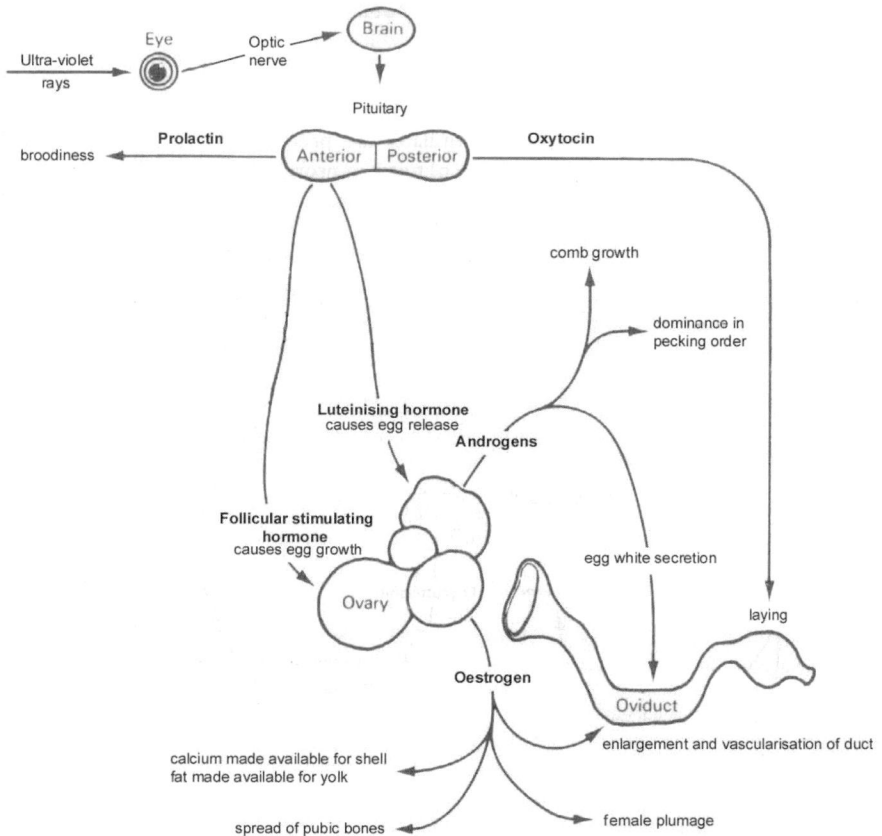

Figure 11 Effect of Hormones on Laying

Ideal Laying Conditions

The laying rate can be maximised by providing ideal feeding and living conditions.

Table 4 Ideal Conditions needed for Laying.

Condition	Requirement
Feed	778 g
Water	0.2 litres/day
Light period	14-16 hours
Light intensity	0.5 lux
Ventilation	12 m^3/hour
Oxygen	at least 11%
Temperature	13-24 C
Relative humidity	50-80 %
Space	1.8 m^2

Such conditions need to be in place during rearing. The optimum day length of 16 hours should be provided in gradual steps to provide maximum general growth as well as ova maturation. The importance of a good supply of water in maintaining egg production is often over-looked.

Other factors which affect egg laying are health, the freedom from disease or parasites and disturbance. This might be unexpected noise, thunder storms, presence of predators etc. In communal situations aggression and the hens position in the pecking order may affect food intake and so lower egg laying.

As hens get older so does their reproductive capacity. A pullet that starts laying in October may well continue without a break until the following winter. After a short winter break she will start again but as she increases in age so does the length of her winter break. A six year old hen would produce half the number of eggs that she laid in her first year. However second year eggs would be larger than those laid as a pullet. Most commercial enterprises get rid of their layers after 16 months, if kept after a moult they would pay their for a further six months. Small producers and back-yarders would probably consider three laying seasons the economic limit.

Candling Eggs

Present day candlers – one battery, one mains operated

Chapter 5

EGG QUALITY

Early Controls: Pre-1950

Before the 1930's egg production was in the hands of small producers or part of a larger farm enterprise. A 'National Mark Scheme' had been established. The aim was to provide a reliable quality of product and it was administered by the government and the Farmer's Union. Its voluntary nature, however, meant that only a small proportion of the market was involved.

During the 1939-45 war foodstuffs including eggs were brought under control of government and rationed. The National Egg Distributors Association was charged with marketing and continued in this role until 1953

In 1949 the Ministry of Agriculture Fisheries and Food published a guide for the testing of egg quality. At this time egg producers were required to sell their eggs through Packing Stations, collections being made from the farm.

The Candling Process

It was at the packing stations that quality was checked, primarily by candling – the shining of a light through the egg in a darkened room. Defects in shell and contents were then apparent, the eggs being graded and the producer being paid accordingly.

Eggs with blood spots, meat spots, broken yolks, a cracked shell and other defects were placed on one side for using in some process or thrown away so that they were not available as food in any form.

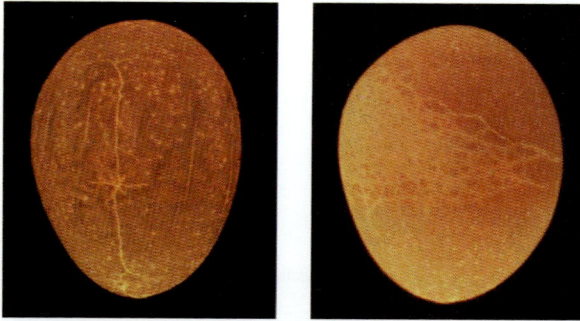

Candled eggs showing shell imperfections

The Eggs Order 1953

This defined two qualities of hen's eggs, first and second.

Table 5 Class A and Class B Eggs Compared

Some of the above faults are due to breeding but many can be prevented by care on the farm. Good husbandry will prevent contamination of shells from dirty litter. Plenty of clean nest boxes, regular egg collection, careful handling and correct storage (20°C) are all self evident measures that were set out as advice to the producer. Other advice included removing the cockerels and broody hens, put nest boxes in shady position, pack eggs broad end up and market quickly. Advice that fifty years later was supported by legislation.

Lion Eggs – The Marketing Board

From 1957-1971 the British Egg Marketing Board, a producer controlled body, took over generic advertising, research and quality control. Any producer with more than 50 hens had to be registered. The mark of quality was a Lion stamped on the egg – this continued until 1968, although it was resurrected in the 90's. The slogan 'Go to work on an Egg' became a National catchphrase.

CLASS A (First Quality)

Excellent internal quality and has three distinct parts: the yolk (visible under candling as a shadow only), a clear translucent white of gelatinous consistency and an outer layer of thin white.

Yolk central.

A small air cell at the broad end of the egg.

CLASS B (Second Quality)

Fair internal quality. The yolk flattening, the two layers of white mingling.

Yolk moving from its central position.

Air cell increasing in size.

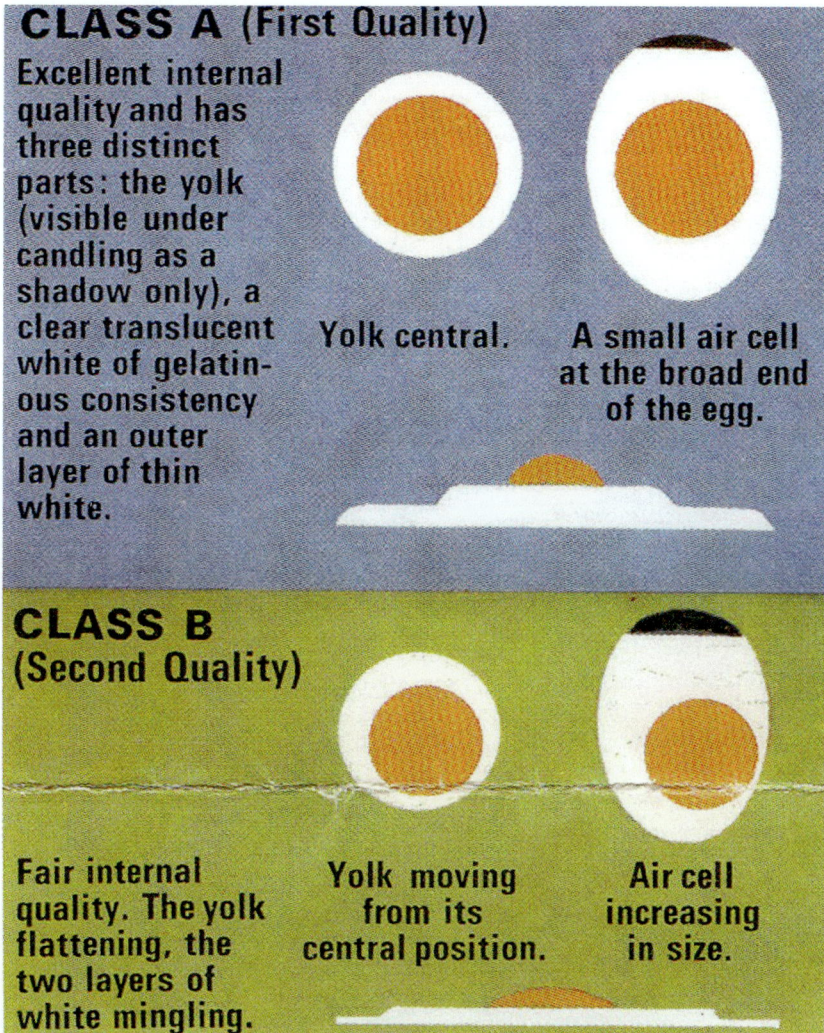

Table 5 A & B Eggs compared

Egg Authority & European Control

Control of the market was not universally popular with producers and a new organisation, the Eggs Authority came into force in 1971. It was concerned with advertising, education and research into diet and health. It was in place when Britain entered the E.E.C. in 1973. Much European legislation was then forced on the U.K. Eggs were graded A B and C, the latter grade were only for use by the food industry. Seven weight grades were mandatory.

Table 6 The Seven Weight Grades

EU Egg Weigher

These proved to be very confusing for shoppers and were later replaced by 4 grades,

| Very Large | 73 g and over | Medium | 53-63 g |
| Large | 63-73 g | Small | 53 g and under |

A good example of things going round in circles!

British Egg Industry Council

This statutory body ceased operation in 1986 and was replaced by the BEIC (British Egg Industry Council). This organisation, independent of government, represented the views of all the areas of the UK egg industry from hatching to marketing. One of its first tasks was to reassure the public after the Salmonella scare of 1988.

It did this by providing a Code of Practice on hygiene and animal welfare that was far more vigorous than that required by U.K. and E.U. law. Hens and the eggs they laid were for the first time traceable. The Lion was brought back as the symbol to denote eggs produced to these standards.

Controls were required at every stage of production from testing the feed to the final handling and storage. Eggs marketed from different from different systems of management had to be clearly indicated eg 'free range', barn eggs, or 'caged'. For each it is necessary to fulfil rigid requirements in terms of space per bird, stocking density, type of litter, access to grass, ventilation, lighting health, hygiene, food and water. By 1998 compulsory vaccination against Salmonella enteritidis was required as well as a 21 day best before date on eggs and egg packs.

Egg Packs

Statutory details are now essential on egg packaging.

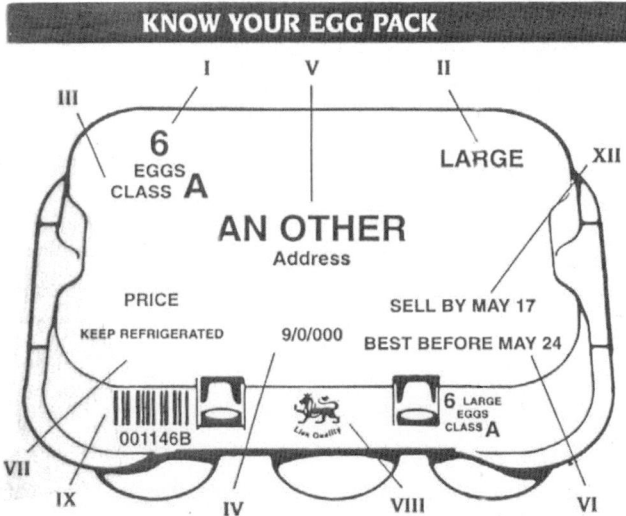

Fig 13 The Egg Box and what it Tells the Consumer

Key

i   quantity
ii   size
iii   quality
iv   reg. packer number (first digit denotes country)
v   packer address
vi   best before date
vii   storage advice
viii   quality mark
ix   retail bar code
x   area of production
xi   production method
xii   sell by date

The recommended *sell by date* is an indication of the last date eggs should be offered for sale to the consumer, after which there remains a reasonable storage period of at least seven days in the home. The best before date corresponds to the end of this storage period. If the laying date is shown on the box it must be shown on the eggs as well.

Production Methods (Systems of Management)
Summary of Regulations

If methods of production are shown on small egg packs, the producers must comply with specific standards of production if these terms are to be used. The permitted terms are:

1. Free Range Eggs
- hens have continuous daytime access to open-air runs
- the ground to which hens have access is mainly covered with vegetation
- the maximum stocking density is not greater than 1,000 hens per hectare of ground available to the hens or one hen per 10 square metres
- the interior of the building must satisfy the conditions specified in the regulations.

2. Semi-Intensive Eggs
- hens have continuous daytime access to open-air runs
- the ground to which hens have access is mainly covered with vegetation
- the maximum stocking density is not greater than 4,000 hens per hectare of ground available to the hens or one hen per 2.5 square metres
- the interior of the building must satisfy the conditions specified in the regulations.

3. Deep Litter Eggs
- the maximum stocking density is not greater than 7 hens per square metre of floor space available to the hens
- at least a third of this floor area is covered with a litter material such as straw, wood shavings, sand or turf
- a sufficiently large part of the floor area available to the hens is used for the collection of bird droppings.

4. Perchery Eggs (includes barn eggs)
- the maximum stocking density is not greater than 25 hens per square metre of floor in that part of the building available to the hens
- the interior of the building is fitted with perches of a length sufficient to ensure at least 15 centimetres of perch space for each hen.

Organic Eggs

One further control on standard is provided by organic eggs. These have to meet the requirements of the Soil Association. The welfare conditions do not vary from those required by the Lion standard. However the egg layers have to have come from organic parents and be fed with organic food – at least 70% of their intake must come from known organic sources.

Present Day Fears

Despite all these controls fears still do exist about egg quality and the supermarkets are proving to be powerful forces in shaping public opinions. Christmas 1999 saw the sale of 'Ready Egg' by Tesco. This contains five pasteurised ready beaten eggs which have no artificial additives or colourings. The fat content has been halved to appeal to weight watchers. Another supermarket chain is banning the sale of eggs from chickens given 'artificial' pigments in their feed. It has long been known that the depth of yolk colour is dependent on pigments obtained in the diet from plant material and grass in free range systems. In cage and barn systems the feed is supplemented with 'artificial colourings, citrana xanthin (E161) and beta-apo8carotenal (E160). But the rich golden yellow yolks so obtained are under threat in spite of the fact that the Lion standard of quality allows for the pigments to be added. Research has shown them to be perfectly safe, and that the public prefer rich orange yolks. The pigments added are those from obtained by free range hens from their diet of plant material and grass. A chemical is a chemical and the same chemical whether it is found in grass, extracted from grass or made in a lab before being added to a poultry feed. One begins to question common understanding of science.

A footnote to this chapter on quality concerns a claim made for eggs from Marans hens, in a national newspaper when the Salmonella crisis was at its height.

It was claimed "the eggs are the only ones in the world totally free of both salmonella and cholesterol. These 'eggmazing', 'eggstonishing' eggs come from France. They were said to have thicker shells, double skins, double protein and be low in cholestrol.

What ever the truth of these claims, we have to be content with normal eggs that have been tried and tested from stock bred from well established breeds.

*W Powell-Owen

W Powell-Owen was a leading personality in the Poultry Club and was President for a record 11 times. He was a great believer in developing utility characteristics, this included eggs for which he established the first standard, accepted by the Poultry Club in 1948.

SECTION 2
THE SHOW EGG

Chapter 6

The Egg Standard

The egg standard for showing was set down much later than the standards for poultry. W. Powell-Owen* used a score card system during the 1930s and 40s which formed the basis of the standard that was accepted by the poultry in 1948. This was modified for the 1982, 4th edition, of the standards on the advice of Max Butler. Further amplification was made by M G V Thompson for the current 5th edition, 1997. The current egg standard is as follows:

The Present Standard

The Poultry Club has authorized the following standard and scale of points for judging eggs.(reproduced by kind permission):

External

Shape: Showing ample breadth, good dome, with greater length than width, the top to be much roomier than the bottom and more curved. The bottom should not be too pointed, and a circular, or even narrow shape is undesirable. The ideal shape is described as an elliptical cone. In outline it is an asymmetrical ellipse or 'Cassinian oval' and a cross-section at any point across the egg's girth is a perfect circle. This description is best shown by the large fowl egg. Pullet eggs are less pointed whereas some breeds of bantams characteristically lay more pointed eggs.

Turkeys, ducks and geese are distinct species and each lay eggs of slightly different shape. Hence, they should be shown in their own classes. Turkey eggs are quite short and conical. Duck eggs are slightly elongated and those of bantam ducks tend to be pointed. Geese lay eggs which are lacking for girth and narrow towards the pointed end.

Size

Mere size is not a deciding point but should be appropriate for the breed and species. A pullet's normal egg when the bird starts to lay is 49.6 g (1¾ oz) and increases quickly to 56.7 g (2 oz), exceeding that after several months of production. There is another increase in the hen egg after the moult. Bantam eggs should not exceed 42.5 g (1½ oz.) Eggs weighing in excess of this should be passed.

Turkey and duck eggs weigh between 70.9 g (2½ oz) and 92.2 g (3¼ oz). Bantam duck eggs should not exceed 63.8 g (2¼ oz). Goose eggs vary with breed. Light geese lay eggs from 141.8 g (5 oz) and heavy breed goose eggs can weigh up to 198.6 g (7 oz).

Shell texture

Smooth, free from lines or bulges, evenly limed, smooth at each end, without roughness, porous parts or lime pimples.

Colour

White, cream, light brown (tinted), brown, mottled or speckled, blue, green, olive and plum. The colour should be even and in the case of mottled or speckled eggs, regular mottles or speckles are preferred. Mottled or speckled eggs are shown according to their ground colour. Where a Breed Club has stated in its standard that a particular colour is required, any variations from this should be penalised.

Freshness, bloom and appearance

Shells to be clean, without dull or stale appearance as befits a new-laid egg. Shell surfaces may be shiny or matt, but should be free from blemishes such as stains and nest marks. Eggs may be washed in preparation but not polished.

In duck eggs the position of the air space can be apparent. This is not considered a fault. Muscovy duck eggs often have a wax cuticle which may be removed.

Matching and uniformity

Eggs forming a plate or exhibit to be uniform in shape, shell texture, size, colour and appearance.

INTERNAL
Yolk

Rich, bright golden-yellow, free from blood streaks or 'meat' spots. Well-rounded and well-raised from the centre of the albumen. One uniform shade. Blastoderm or germ spot not discoloured and there should be no sign of embryo development.

Albumen

This is clear with no signs of blood spots or cloudiness and preferably with no tint of colour. It is of dense substance, particularly around the yolk and the differentiation between this thick albumen and the thin outer should be distinct. Waterfowl albumen must be clear, it is also more viscous and distinct than the hen's albumen and for these reasons water-fowl contents should be exhibited in classes separate from large fowl and bantam.

Chalazae

Each chalaza to resemble a thick cord of white albumen opposite each other and attached to the yolk, keeping it to the centre of the inner albumen. Free of blood and 'meat' spots.

Airspace

Small, about 1.5 cm (½ in) diameter (1 cm (³⁄₁₆ in) bantams), the membrane adhering to the shell. It should be placed at the broad (dome) end ideally just to one side.

Freshness

Indicated by small, taut airspace and unwrinkled top surface of yolk which should be raised and not lacking in height. A stale albumen lacks differentiation and is watery and runny.

Scale of points

External		Internal Points	
Shape	25	Yolk	30
Size	15	Albumen	30
Shell texture	20	Chalazae	10
Colour	20	Airspace	10
Freshness, bloom and appearance	20	Freshness	20
	100*		100

Serious defects
(for which eggs should be passed)

More than one yolk. Staleness. Polished or over-prepared shells. Overweight in bantam eggs including contents classes. A developing embryo as shown by a 'halo' around the germ spot. Excessive blood streaks and 'meat' spots.

Disqualification

Addition of colouring to shells. Artificial polish or colouring would amount to disqualification and a report to the Poultry Club of Great Britain.

*May be maximum for each egg, or for a plate of eggs, whatever the number. Add 5 points more for each egg for matching and uniformity

Chapter 7

EGG SHAPE

S hape is everything to most egg judges. A well-matched plate of three eggs for size and colour will rarely succeed if their shape is poor. Shape to an egg judge is like type to a poultry judge; 'type gives the breed, colour and variety'.

In both cases, however, the points as allocated in the Standard do not really justify such emphasis. For eggs, 25 points are awarded to shape, 15 to size, shell texture, colour and freshness score 20 points each. This indicates that shape should be given only marginally more importance when judging.

What is a good shaped egg? Is it just a matter of preference? For guidance one can consult the Poultry Club's Book of Standards (published by Blackwells) On pages 360 and 361 several eggs are illustrated but variations are apparent, either in reality or because of the angle at

EXTREMES OF EGG SHAPE

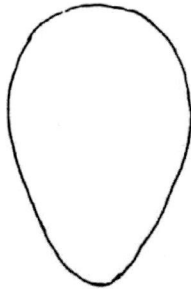

Guillemot Egg   Owl Egg

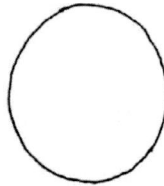

Fig 14 Two extremes of egg shape-the owl and the guillemot

which the photograph was taken. Only in a diagram can the 'perfect' egg be shown.

If other birds' eggs are examined it is obvious that a wide range of shapes exist. The egg is the home of the developing chick and as such, it must survive the bumps and bangs of brooding, hence a strong shell. It must not be eaten by predators, hence camouflage and it must remain in the nest, hence shape. A pointed egg will roll in a circle rather than roll away, so not surprisingly the Guillemot which nests on cliff edges has an egg of this shape. Contrast this with the round egg of an owl whose clutch lives within a firm, confined space. Round eggs in this situation will move freely during turning by the mother, long, pointed eggs would tend to jam and break.

What Controls Egg shape?

Such extremes exist as the result of evolution by selection on the birds' inherited material, their genes. Most books on genetics give the heritability of egg shape as being in the order of 40 – 50%. This means that about half the influence on shape is by way of genes, the rest is the result of environmental factors. Such factors could be the amount of food or water, the fact it was the first in a string of eggs or that stress occurred during laying. A bird, therefore is likely to lay a particular shape of egg, but there is no guarantee that it will do so.

Since egg shape is influenced by several pairs of genes, rather than a single pair, a wide range of shapes result. Shape, as we have seen, varies from the very round to the very long.

So back to my original question, what is the perfect shape for a chicken egg?

In nature chickens lay neither on rock ledges or in a hole. Perhaps there are other selective influences at work on their genes. I contend that there are two. As a chick develops it turns within the egg, turning is aided by ample girth. At about 12 days incubation the chick instinctively gets its head to the highest point in the egg. This, if the egg is broader at one end than the other, will be the end where the air sac lies. A long narrow egg in a nest will not naturally lie with the air sac end higher than the other so there is a fair chance the beak will not penetrate the air sac and the chick will fail to hatch.

During incubation of the chick, temperature and rate of water loss by evaporation are crucial factors for successful hatching. Heat exchange and water loss are affected by the ratio that exists between shell surface area and volume of contents. A long thin egg will be more susceptible to external changes than a broad, round one, so will be less likely to hatch. For a chicken then I suggest these are the reasons why a broad, rounded shape is beneficial for successful breeding and why shape should be re-garded as the most significant of the judging features.

The Mathematical Ideal

Further support for the broad, rounded egg comes from Romanoff and Romanoff in their classic book on the subject entitled *The Avian Egg*. They suggest the ideal shape for a chicken is a 'Cassinian oval' – an asymmetrical elipse – with one end blunter than the other. They devised a mathematical way of quantifying egg shape. They measure the length of the egg and its width two thirds down from the broad end, which should be the widest point.

These measurements can be used to work out the shape index in the following way:

$$\frac{\textbf{breadth}}{\textbf{length}} \times 100 = \textbf{shape index}$$

According to their research the ideal measurements for a hen's egg is

$$\frac{\textbf{4.25 cm breadth}}{\textbf{5.75 length}} \textbf{ x 100 = an index of 74}$$

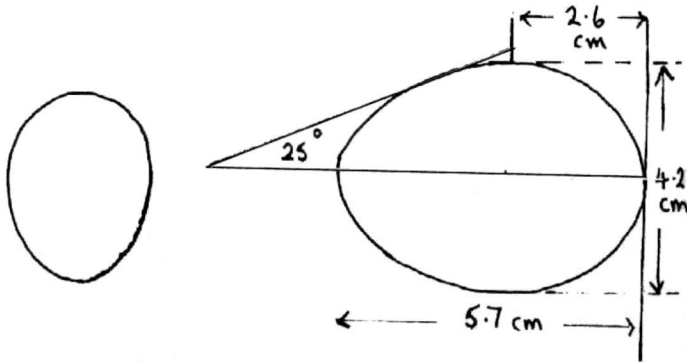

**Fig 15 Diagram illustrating the measurements needed
to calculate the Egg Index**

Judging the Shape

The skill of judging an egg for shape is to hold the ideal in one's mind as the various eggs shapes presented by the exhibitors pass before one's eyes. A vital element in recognising the ideal shape is provided by the outline. This is much easier to see against an even background hence the green or black cloths used by judges.

If other backgrounds are use the judging becomes more difficult. For example the outline is blurred by shavings and sawdust. Moreover, just as chickens cannot be satisfactorily judged through the wire so eggs cannot be judged on a plate. Outline is also blurred by any blotches and speckles of colour, so judges must take great care when judging mottled egg classes. Some judges use a wooden rod against which the egg curve can be assessed.

Fig 17 shows various eggs contained within wood rules and set against a dark background which illustrates the importance of recognising outline to gauge shape. Also included for each egg is its shape index value. Any value between 72 and 76 is near the shape for show winners.

It must be remembered, however, that index alone does not totally represent shape. Either side of the measured points the shell may bulge as shown by egg A (lower right side). Or the sides of the egg might not be even in curve, egg M. A good way to check for these uneven curves is to hold the egg broad end down and rotate it while looking at the pointed end., any bulges soon catch the light and disclose their presence.

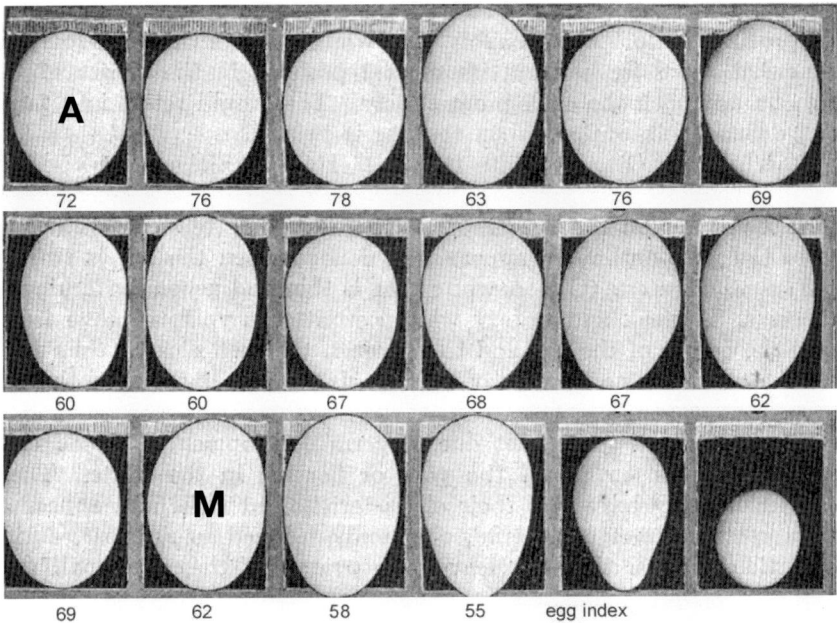

Fig 16 Eggs within wooden frames to illustrate shape variation and the egg index

Egg Shape in Different Classes of Poultry

Variation of shape is found within poultry. Geese lay rather elongated eggs, Romanoff and Romanoff give 70 as their shape index. Most ducks lay eggs just slightly longer than chickens e.g. Pekins, but Runners and Muscovies average the ideal 74. Bantam ducks being nearer the ancestral Mallard, have an index of 71. Turkey eggs tend to taper quite sharply towards the pointed end giving them a conical shape. Guinea fowl eggs have a similar conical appearance but are much shorter which gives them an index of 76. Silkie fowl tend towards the oval. It is not surprising then that the Poultry Club suggest that eggs from different categories of poultry should be shown in separate classes.

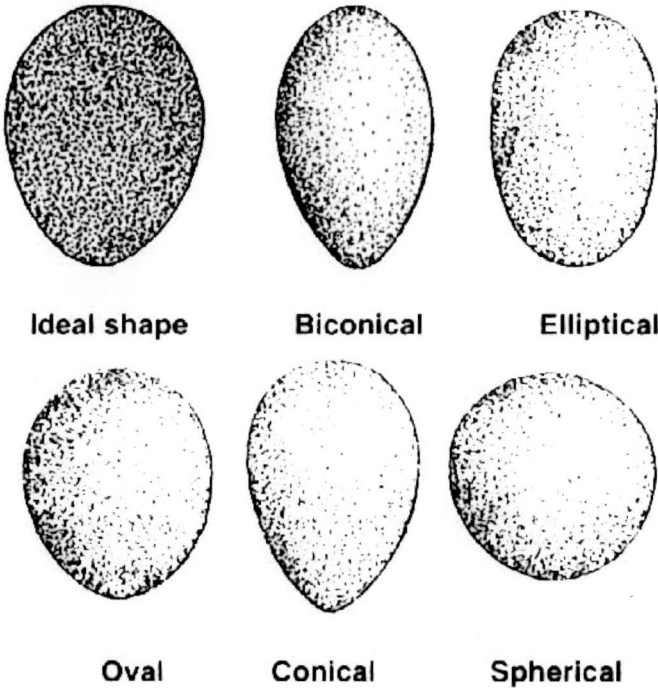

Ideal shape   Biconical   Elliptical

Oval   Conical   Spherical

Fig 17 The Ideal Egg Shape and some common variations

Chapter 8

EGG SIZE

In the 1982 revision of the egg standard size was demoted in importance from 20 points to 15. Shape which increased its points allocation to 25 was considered more important. Shape makes the show egg, colour and size provides the variety.

Egg Size in different Types of Poultry

What causes this variety? First and foremost different species. Ducks, geese, turkeys and chicken all have a different mean size. In the 1997 Poultry Club egg standard, ideal weights are laid down. Turkey and duck eggs mean weight lies between 71 g (2½ ozs) and 92 g (3¼ ozs). Bantam duck eggs should not exceed 64 g (2¼ ozs). Goose eggs vary considerably. Light geese such as the Chinese lay eggs of 142 g (5 ozs), while heavy breeds like the Emden and Toulouse can weigh up to 197 g (7 ozs). The standard weight of the hen egg according to Romanoff is 58 g (2 ozs). Miniatures being between a quarter and fifth the size of their large fowl counterparts are required to lay an egg not more than 42.5 g (1½ oz). Above this and they will be passed in competition. True bantam eggs rarely reach this limit averaging 29 g (1 oz).

Egg Size in Different Breeds of Poultry

Variations from these specific standard weights occur with breed. Heavy breeds generally lay larger eggs but there are exceptions. The Welsummers, a bred in which there has been selection for the flowerpot brown egg, now lays a big egg despite its comparatively light body weight. Some of the light breeds such as the Minorcas lay a large egg as Table 7 shows.

Table 7 Comparative weight of eggs of some representative breeds of fowl

Breeds of Fowl	Average egg weight (grams)
Dark Brahma	68.9
Plymouth Rock	63.9
Minorca	63.9
Andalusian	63.4
Light Brahma	62.0
Ancona	61.4
Orpington	60.1
Rhode Island Red	59.3
Leghorn	58.1
Wyandotte	58.1
Hamburg	57.1
Japanese Bantam	30.8
Jungle Fowl	29.1

Table 8 Change in egg weight during Domestication

Ancester Species	Modern Species	Egg Weights (Grams)	% Increase
Jungle Fowl	Dark Brahma	40 g/68 g	70.0%
Canada goose	Embden	135 g/215 g	59.3%
Mallard	Pekin duck	57 g/83 g	45.5%
Wild Turkey	White Holland	75 g/85 g	13.3%

NOTE: The size is influenced by the environment and genetics which is measured by heritability. (See *Genetics & Evolution of the Domestic Fowl*, Lewis Stevens, Cambridge, 1991.

Such breed differences have their origins in genetics. Individuals within a breed have egg weights that produce a 'normal' distribution curve around the mean so it is likely that several genes are involved in controlling size. Some work in the 1920s indicated that small egg is 'dominant' over the large. Further evidence for this is the fact that if large egg size is not selected in breeding birds, succeeding generations have a tendency to lay smaller and smaller eggs. They revert to the egg size of their natural ancestors. Table 8 shows how effective selection for egg size has been during domestication.

Selection in game birds has paid little attention to egg size and as a result their eggs are rather small in relation to their body weight. Extreme inbreeding has also been shown to reduce the size of the egg.

To date there is no confirmed scientific evidence as to the influence of one parent over the other. Most believe that both parents contribute equally to the egg weight of their offspring.

Egg Size in Different Individuals of the same breed

As well as variations between breeds there is variation of individuals within the breed. Genes are significant in these variations but the environmental influences upon each of the individuals plays an important part. Different living conditions and diet during development will result in birds with differing characteristics. Body size and weight has long been recognised as influencing egg size. The heavier the body the larger the egg. When judging birds for utility, body size is taken into account. To quote Powell-Owen 'The size of frame is associated with the size of egg. Select the bird with wide back, ample space between the legs, full abdomen and wide pelvic arches.'

The food of a bird, especially the level of protein in the diet, influences egg laying but it affects the rate and number of eggs more than the size. Calcium is critical, when calcium phosphate is the only source egg weight declines. Vitamin D is also significant. Some of the drugs used for worming can cause a temporary decline in size.

The health of the bird affects egg laying, an ill bird stops laying almost immediately. Feeling out of sorts means that the bird will not be feeding and so the rate of laying will be lowered but not the size of the egg.

Climate, like food and health, affects the number of eggs laid rather than size, although there is some evidence to suggest smaller eggs are laid when the temperature is abnormally high. Light, which is significant in controlling the timing of egg laying, has no apparent affect on egg weight.

Confinement in cages enables close control of diet and it restricts physical activity which results in an improvement in the efficiency of food utilization. There is no evidence that egg size increases in this situation, only that more eggs are laid.

Then there are the effects of time. Age is significant. A pullet egg at the start of lay is about 49 g (1-1½ oz), this gradually rises to about 57 g (2 ozs). As laying season follows laying season the average weight varies, see Table 9.

Table 9 Changes in average Egg Weight with Age

Laying Year	Leghorn	Plymouth Rock
1	53.2 g	59.5 g
2	56.8 g	60.2 g
3	56.5 g	64.1 g
4	56.0 g	61.8 g
5	54.1 g	58.9 g
6	53.7 g	59.2 g
7	52.8 g	58.8 g

Egg size will vary according to the season, the largest eggs are laid in Spring. The size of eggs varies with the time of day it is laid. This is linked to the cycle of egg laying. An egg laid at the start of a cycle is laid in the morning and is the heaviest. Succeeding eggs are smaller and as the period between laying is just over 24 hours they are laid later the next day. These egg laying cycles which have their origins in the laying of clutches of eggs vary in duration. The longer the cycle the smaller the average decrease in weight.

Optimum Size for the Show Egg

In most livestock show classes, if all other things are equal, the largest animal wins. Should this be so with eggs?

To answer this question two things should be borne in mind. Firstly the chicken can only process so much food in 24 hours. At the end of the month is it better to have thirty grade 2 eggs or twenty grade 1? Some would argue, too, that producing large eggs puts a strain on the system and shortens the egg laying potential. Secondly, the egg is the next generation of hens. Within its shell is the yolk and albumen that contains the necessary nutrients for the chick development. 'A good sized egg means a good strong chick'. Quite true, but the balance must be right. A large egg with limited contents will produce a large but weak chick.

During development the egg is warmed and cooled, oxygen enters and waste gases exit through the shell. The speed at which these processes occur is related to the surface area of the egg. The amount of heat and oxygen needed varies with the volume of the egg. It follows then that the ratio between the surface area and volume of the egg is crucial. A small egg has a large surface to volume, lots of area for gas exchange and heat gain. As the egg gets bigger the area to volume ratio becomes less, a large body has comparatively less surface through which to exchange gas and heat up.

I contend that for each species there is an optimum size. A size which balances the quantity and quality of yolk and albumen with the ratio of shell surface to volume so ensuring the development of a strong chick. When judging I would give maximum points for the ideal 58 g-70 g (2-2 ozs) and deduct points for eggs that were smaller or larger.

Large and bantam egg in a pair class

Fig 18 Egg size in different species (to scale)
1. Aepyornis – Elephant bird, 2. Ostrich,
3. Chicken, 4. Humming bird

Chapter 9

EGG SHELL

Function of the Shell

The function of the shell is to provide a safe incubation chamber for the developing chick. It must be strong to withstand the weeks of sitting yet porous to enable air to reach the chick. There is no doubting its importance to the well being of the chick and it is allocated 20 points in the standard – the same as colour and freshness. An egg with a poor shell is unlikely to survive the rigours of brooding.

Evolution of the Shell

The first 'land' eggs laid by the reptiles were encased in flexible shells. Some reptiles guarded their eggs, others had nothing more to do with them but none incubated the eggs. Heat for incubation was provided by the sun beating down on the sand e.g. turtles; or from the rotting of vegetation, e.g. grass snakes. Birds, however, sat on their eggs which provided them with heat from their bodies. The eggs were laid in a variety of nests, some complex structures, some just scrapes in the ground. The contact of parent birds entering and leaving the nest together with the movement of the eggs against each other required a stronger exterior than that of the reptiles – a hard shell.

This shell needed to be strong throughout incubation, while providing the embryo with calcium and magnesium for bone development. It also had to allow the exchange of gases between the developing chick and the outside. So holes in the shell there had to be, but how to prevent the entry of water in aquatic nests and harmful micro-organisms? Muscovy ducks provide a wax covering that repels water as the clutch is being laid. Once incubation starts the constant shuffling of eggs rubs it off allowing more air to exchange. Harmful micro-organisms such as the salmonella bacterium can enter eggs via the shell pores although these are usually blocked when the cuticle is deposited during the final laying process. Most evidence suggests that bacterial entry is via flaws in the shell such as hair cracks which are produced by the uneven deposition of the shell or from physical knocks to the egg.

The present egg shell is thus a compromise structure, developed over the years by evolutionary selection in response to the conflicting demands. As with all new evolutionary structures it was built on past structures – a shell designed from scratch could have been very different. Thus the shell that was developed was based upon the membranes already laid down and used the reptile egg laying apparatus. It was a specialisation of part of this, the shell gland, that enabled the bird to produce the thick shell.

Shell Formation

The initial deposition of calcium salts occurs in the tubular shell gland – an area of the oviduct between the isthmus and shell gland pouch. The calcium salts are deposited onto the ends of the membrane fibres. These areas then attract further salts whose aggregation form the start of the mammillary bodies or knobs. It is important that these bodies are evenly distributed on the membrane and that the bonding between the organic membrane and the inorganic salts is good in order to provide a firm base for subsequent shell development. It is known that excess sulphates in the membrane can cause weak bonding.

Subsequent shell construction occurs in the 20 hours that the developing egg remains in the shell gland pouch. Calcium salts are deposited on the mammillary 'knobs' and further mineralisation produces the cone or palisade layer. The salts are deposited into the previously secreted collagen matrix, the final mix being 95% calcium salts and 5% protein.

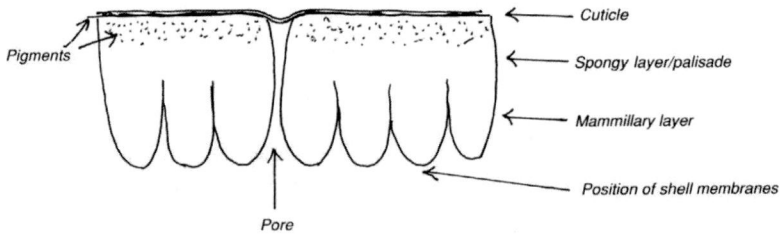

Fig 19 Egg Shell Structure

Some areas remain free from salts and these form a network of spaces linking the gaps of the mammillary cones to pores in outer shell surface. Varying pH, the amount and purity of the salts all effect the rate and quality of shell production. Before the shell has fully encased the membranes water is 'pumped' into the albumen, increasing its weight by up to 15 g. The last layer of calcium to be added forms crystals that are aligned vertically forming the 'vertical crystal layer' upon which the cuticle is secreted. This is produced by the basal cells of the pouch wall and is constructed of proteins like the egg membranes and includes porphyrins which become florescent in U.V. It surrounds the shell including the pore openings. Any colour pigment characteristic of the breed is also produced at this stage. This then completes the shell construction.

Factors Affecting Shell Strength

Throughout the process any imbalance of materials or change in the conditions of the pouch caused by external factors such as stress can affect the laying down of materials. This can lead to fissures or areas of weakness that will affect the strength and durability of the shell. Thus a genetic 'blue print' for the perfect shell can be considerably modified during implementation.

There are those who would argue that intensive methods of poultry rearing and selection for uniformity of size and colour mitigate against the original purpose of the shell which was to provide the ideal embryonic chamber. Certainly poultry kept for providing hatching eggs should be given the 'best' possible husbandry and environment. They must have a high protein ration, calcium and phosphorus with Vitamin D that aids their metabolism, together with a trace of manganese for shell thickness.

What is the effect of fissures within the shell? It is commonly believed that they weaken the strength of the shell. This is partly true as they lower the shell's resistance to cracking. Extensive cracking will weaken the egg but more importantly enable micro-organisms to enter during storage as well as accelerating evaporation. Both cause deterioration of the contents for both eating and incubation.

The oval elliptical shape of the egg provides a 'built in strength' for when pressures are applied the forces are such that they work to produce a resistance to breaking. Given a non cracked shell it is quite difficult to crush an egg between the fingers or in the palm of one's hand.

Examples of faulty Shell Formation

Bulges

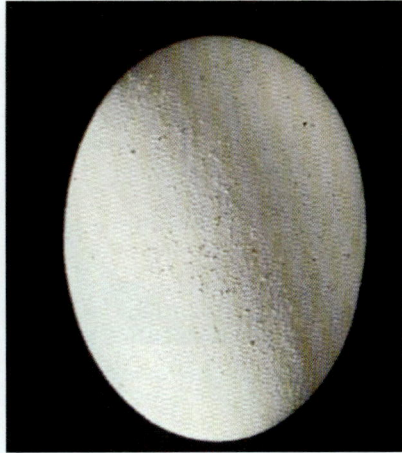

Pimples

Measuring Shell Strength

Several tests are used to gauge the strength of objects most of which in-volve applying known forces, either by compression between two plates or by impact e.g. the dropping of a known weight from a standard height. Most damage occurs to eggs by impact during transport. It is the degree of spread of cracks from the point of impact that is significant in causing 'ultimate stress failure'(UTS). The ability to resist UTS is called fracture toughness. This depends upon the shell's ability to resist stress forces and to absorb the energy applied by the impact. Experimental evidence demonstrates that there is a direct relationship between the degree of fusion between the adjacent palisade column and the number of mammillae per unit area. Any fissures will also lower the shell fracture toughness. There is little strength in the mammillary layer and cuticle.

Thus the strength of the shell is the result of the shape of the egg, an ideal deposition of chemicals, its porosity and thickness.

Shell Thickness

Thickness can vary considerably within the clutch and season. Shell thickness can be determined indirectly by estimating the amount of shell from weighing the egg in air and in water.

$$\text{Shell amount} = \frac{\text{weight of egg in air}}{\text{weight of egg in air} - \text{weight of egg in water}}$$

However, observations of the shell exterior can provide clues as to a shell's strength. The following list includes some of the most obvious features and are ones that are included as undesirable or defects on the Poultry Club egg standard.

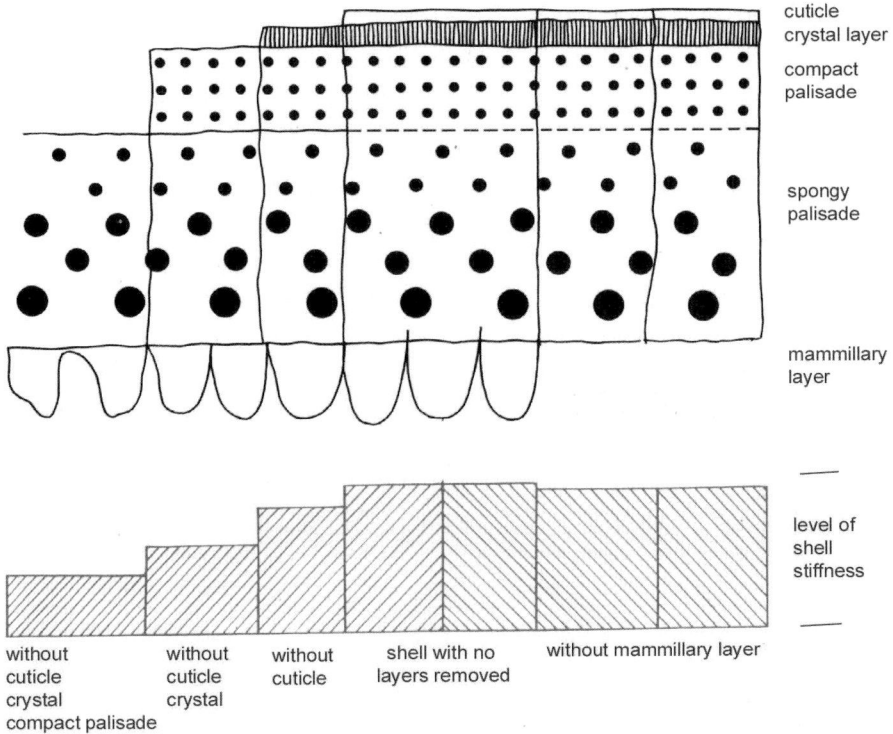

Fig 21 Diagram showing the Contribution of the Shell Layers to Strength.

Shell Defects that affect Shell Strength

Calcium splash – this is a chalky deposit on the egg surface. On brown eggs this gives a pink/lilac appearance. It is due to an ionic imbalance in the shell pouch gland caused possibly by stress.

Slab sided egg – this uneven shape is due to two eggs being in the pouch gland together.

Soft shelled egg – little mineral deposition because of an imperfect foundation in the mammillary layer. Cause stress. Stress can also accelerate the passage of the egg so no deposition occurs.

Equatorial bulge – the result of a repair to a break in the shell in the pouch due to excessive muscle contraction of the walls. Checks can occur elsewhere on the egg.

Corrugated eggs – the result of displaced mammillary 'knobs' on the membrane due possibly to abnormal water absorption. This can be prompted by Infectious Bronchitis.

Translucent shell spots – these can be seen when fresh eggs are candled, they also become obvious with age. Commonly considered to be due to uneven shell thickness but in fact the result of irregular fusion in the mammillary columns. The gaps so formed attract water from the albumen.

Shell accretions – calcareous pimples produced from malfunctioning glands. Having failed to release Calcium during mineralisation they deposit it all at once just before egg laying.

Stretch marks – shallow grooves running from the dome of the egg down the sides due to muscle action on the soft cuticle during the final expulsion.

Wrinkles at narrow end – ineffective mineralisation due to shortage of calcium supply. Can be related to the position of the egg in the clutch (sequence of laying).

Chapter 10

COLOUR IN EGGS

Colour in eggs is immediately obvious, even to the layman. Plates of dark brown eggs and pure white eggs gain admiration and the 'amateur' judge will often be influenced by such colour. At shows green and blue eggs provoke more comment from the general public than do any of the other exhibits.

Striking though colour is, it is given only 20 points on the judging scale. Some might say that 20 points is too much, for colour in poultry eggs is arbitrary and selected by Man. Shell colour does not affect the shelf life or taste of an egg, despite what some advertisements might state! Nor does the colour affect incubation success.

Shape, shell and freshness all have a direct bearing on hatching so it could be argued that these features should have more points allocated to them e.g. Shape 25: Shell 25: Freshness 25: Size 15: Colour 10. However, there are breeds of chicken where egg colour is selected for to satisfy a market demand – for example: Barnvelders, Marans and Welsummers. Their breeders would not wish to see a points standard that gave only 10 points for colour.

No colour is any better than another. The judge is looking for colour evenly applied over the whole egg. Nest marks or scratches on the egg surface, will all detract from the overall effect, as will translucent blotches.

The shell of a fresh egg has a 'bloom' that dulls with age, adversely affecting its colour. In a plate of six or three colour matching between eggs becomes significant.

The shade of colour is only important in particular classes. For example, in a white egg class pure white is required. Cream eggs in this case would be penalised. In a brown-mottled class, two factors are important, the depth of ground colour and the even appearance of the mottling.

Breed classes have special requirements. Marans are known for a brown egg, so the deeper the colour the better. Welsummers, ideally, should be flower pot red and the Croads a plum colour. Those nearest the standard get greater credit. Araucana classes demonstrate the blue/green colour. My personal preference is for a blue hue, but the breed club allows all shades of blue/green but not khaki as this indicates crossing with a brown egg breed.

What gives the Egg its Colour?

Mention of the Araucana leads one to comment about what colour is and where in the shell it is found. Colour in eggs is the result of pigments called ooporphyrins: uroprophyrin, coproporphyrin and protoprophyrin all have been isolated from brown eggs, and biliverdin from blue/green eggs. Some of the pigment does get into the shell but most is secreted with, and remains in the cuticle. The exception is the blue/green colour of the Araucana. This is deposited throughout the shell and is clearly visible on the inside when the egg is cracked open.

The depth of colour can vary within a clutch and does vary with season and age of the hen. This in turn must be related to the bird's ability to make the pigment and the availability of the raw materials. Patches of colour and mottling can remain quite constant for a particular bird. This could be the result of having clusters of pigment producing cells in the pouch wall. There is no evidence that a particular pattern is inherited, but the general colour is the result of interacting genes. Artificial selection for egg colour is effective over a few generations.

Genetics of egg colour

The jungle fowl lays a white egg and thus white can be considered the wild type. It was thought that the white egg colour was linked genetically with white ear lobes. While the majority of white egg layers do have white lobes and brown egg layers red lobes a link is not proved. Mutations from the wild type have produced colour variantions The blue colour of the Araucana egg shell is one such - a dominant gene O which produces oocyan in the egg shell. This occurs on the same chromosome as the gene pea comb so some thought that they were always linked. Other colours as has been said are the result of pigments in the cuticle that are added to the surface of the egg shell. The nature of these cuticle pigments are under the control of at least two genes, one of which acts as an inhibitor. This gives a wide range of colour shade and explains why colour depth can be **selected for over generations. As a rule of thumb: White shell cross brown shell gi**ves intermediate colours. The range of khaki and olive eggs, now popular on the show bench are the result the interaction between the basic blue shell and varying shades of brown pigment in the cuticle. The genes that interact to produce these colours are shown below.

Table illustrating genetics of shell colour of poultry

Shell phenotype	Shell colour oocyan genes	Brown Pigment genes	Brown pigment Inhibitor genes
white	+ +	no	yes
blue	Oo or OO	no	yes
brown	++	yes	no
green	Oo or OO	yes	no

Show Penalties

Uneven colouring is penalised in the egg standard and can be stress related. A dullness in colour – lack of 'bloom' comes with age as the pigments in the cuticle fade and discolour. (see page 72)

The application of polish or artificial colouring to 'freshen' up an ageing egg for show results in disqualification and a report to the ruling body of the Poultry Club. The cuticle when broken by bird's feet in nest boxes develops scratch marks, penalised on the show bench but essential during brooding to expose the pores in shell which enable gas exchange.

The Nature of the Cuticle

The nature of the cuticle varies. In the Ostrich it is very hard, but in the Muscovy duck egg it has a waxy texture. If rubbed with wire wool the cuticle is removed, together with its colour. Further rubbing will then remove the remaining pigment in the outer palisade layer. The brown pigment is literally skin deep.

Colour in Eggs other than Poultry

Reptile eggs are white, so too are most waterfowl eggs and this colour is regarded as the primitive type. Eggs laid in holes, such as Owls', are white, as are pigeon eggs. Both could be said to be out of reach of predators so camouflage is not really required.

Compare these with the remarkable camouflage obtained by birds such as the Terns, whose sandy-blotched eggs merge with the beach. Not all colours are of survival value, however. The blue of the Hedge Sparrow seems rather obvious and why the brown speckles on the Robin or Turkey?

EGG CLASS COLOUR GUIDE

Dark Brown

Light Brown

Cream

White

Blue/Green

Show Classification based on Colour

In poultry four main colours are recognised; brown, tinted, white and blue/green. These are associated with particular breeds; Mediterranean races lay white eggs, Araucanas blue, Barnevelders brown and Sussex and Australorps tinted. Oddly, Americans favour white eggs to eat while the British have grown to prefer brown. Benjamin in 1920 identified

50 shades of colour though he rationalised this to 17 for practical purposes.

In most egg shows there are classes for four to six colours. But even with these few there is confusion, as to what 'shade' should be shown in which class. A classification, such as, brown, tinted, white, A.O.C. (Any Other Colour) for instance, would be likely to produce the following problems: Light brown eggs in both brown and tinted classes, some might even get into A.O.C., and cream eggs in the white class. Some shows allow judges to re-classify the exhibits, so that like competes with like, making the competition a little more fair.

My personal preference for egg classification does away with the general term 'tinted', although I recognise it is widely used in the Poultry Book of Standards. I like the terms, dark brown, light brown (tinted), blue/green, cream and white. Such a five colour system does away with the need for the A.O.C. class. The illustration of paint colours represents my subjective view of which colour goes into the categories I have listed above. Of course, at larger shows particular breed clubs may have separate classes and mottled may stand as a class on its own.

The Poultry Club revised it's rules for showing eggs in 2012 stating that **Tinted** would no longer be recognised as a colour for egg shows. They also ruled that judges should not reclassify exhibits once tabled.

Chapter 11

EGG CONTENTS

To the surprise of many of the general public egg contents are not taken into account when judging most classes of eggs. If the judge thinks the contents are stale then Poultry Club rules allow for them to be inspected but otherwise nothing is known about the contents of the Champion eggs. To the public, eating the egg, the contents are the only important part, the yolk colour in particular. But to the breeder requiring the egg to incubate, it is the size, shape, shell texture, freshness and colour that are significant. These features the breeder can identify and select for, the yolk and albumen remain unseen in the hatching egg unless candled for hair cracks and blood spots.

Most, but not all, shows do now have classes for egg contents. Some have a combined class. The externals are judged first and points given, the egg is then cracked open and points given for internals. The winner is the egg with the highest combined total.

Table 10 The Main Nutrients of the Egg

Constituent	Amount
Water	38.0 g
Protein	6.8 g
Lipids (fats)	6.1 g
Carbohydrates (sugars)	0.5 g
Iron	1.1 mg
Vitamin A (carotene)	77.0 ug
Vitamin D	0.1 ug
Vitamin B1 (riboflavin)	0.3 mg
Kilo joules (energy)	337

Chemical Composition

What makes a 'good' contents. First and foremost the contents are the stored food material for the developing chick. Table 10 shows the main nutrients in an egg.

Other nutrients include small quantities of calcium, thiamin, nicotinic acid and vitamin B12. For non meat eaters eggs are an important source of this vitamin as it is not present in plant foods. Note that there is only a trace of carbohydrate, so eggs are not much use to the marathon runner, and no vitamin C. The protein present contains all the essential amino acids needed for the growing chick. The fat or lipids is emulsified and is readily utilised for providing energy. The egg is an important part of a balanced diet and it is no wonder that there are so many recipes that make use of eggs.

Waterfowl eggs contain less water and more fat than other poultry. This is probably due to the embryo requiring more heat energy as the nests are near water and so are cooler. They also have a slightly higher body temperature.

Table 11 Compares the contents of eggs from various types of poultry

Constituent	Poultry Type			
	Chicken	Turkey	Duck	Goose
Water	73.6%	73.7%	69.7%	70.6%
Protein	12.8%	13.1%	13.7%	14.0%
Lipids (fats)	11.8%	11.7%	14.4%	13.0%
Carbohydrates	1.0%	0.7%	1.2%	1.2%
Inorganic matter	0.8%	0.8%	1.0%	1.2%

Composition of the Yolk

The yolk as explained in Chapter 2 surrounds the germinal disc and is deposited in layers that relate to the feeding activity of the hen. The yellow yolk layers have the higher percentage of proteins (15%) and lipids (36%) and less water (45%) than the white yolk. The precise amounts are difficult to determine because of the closeness of the layers which prevents accurate analysis. 95% of the total yolk is in the yellow layers which is laid down very rapidly. Minerals such as phosphorus, calcium, magnesium, potassium and sodium are present in small but significant amounts. The phospholipid, lecithin, which contains phosphorus is the most abundant.

Colour of the Yolk

As stated before it is the colour of the yolk that is important for the consumer. The colour is the result of pigments, lipochromes which are soluble in fat e.g.carotenoids and xanthophylls and lyochromes which are soluble in water e.g. ovoflavin. All are present in plant leaves and grass which explains the well coloured yolks of hens given free range. If an excess of a particular pigment is available the yolk can take on exaggerated colours which are not acceptable. Experiments that involved the feeding of hens with green and red dyes resulted in green or reddish yolks. This and other evidence supports the view that the nature of the yolk is the result of many environmental influences. Inheritance plays little part. Intensively fed poultry have in their rations extracted pigments that substitute for natural green stuffs.

What is the perfect colour? In the last resort this must depend upon personal taste but market research indicates certain common preferences. The yolk should be of even coloration with no blotches or granular areas. It should be bright and yellow. Too much orange is not popular, particularly if the white has taken on a green tinge that often happens when the yolk is very orange. In the egg industry the colour of the yolk is measured by the *Roche Colour Fan*.

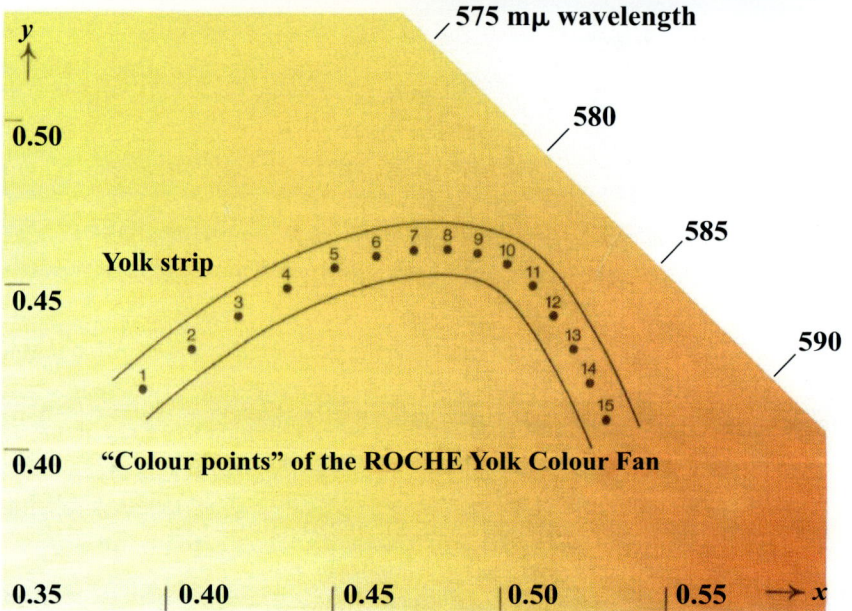

Location of the "colour points" of the ROCHE Yolk Colour Fan
in the "yolk strip"

This is based upon objective measurements of reflected wavelengths light from yolks. The range of colour is from pale yellow to orange-red and fifteen points in this gradation have been defined and reproduced on card strips. This is then the standard scale of colour used throughout the world. Most consumer choice lies between points 8 and 12. When using the fan the yolks should be judged against a dark background, ideally in diffuse daylight since direct light can cause reflections.

Yolk Faults – Blood and Meat Spots

The one other thing that the consumer does not like are the presence of blood or 'meat' spots in the contents. These show up when the egg is candled so eggs with this fault should not appear in the shops. Sometimes the yolk appears smeared with blood, at other times the blood is contained as a spot or two. Blood can occur within the albumen but the most

Faulty contents

common occurrence there are dull pink little lumps – 'meat spots'. They are particularly associated with the chalazae. The blood is released when the blood vessels in the follicle wall that supply the ova briefly rupture. If this occurs just before release then the blood appears on the surface of the yolk. If rupture was early on in the ova's development the blood tends to clump and undergo some degeneration. This is then engulfed by the albumen as it is laid down. Thus these 'meat spots' are of similar origin to blood spots. They are not, as is often quoted, pieces of oviduct that have become detached and enclosed within the albumen. Older textbooks relate the bleeding to the time of egg release from the follicle and indicate that stress plays a part. There is no evidence for this.

What makes a bird produce blood spots is unclear. There may be an inherited tendency, but the environment, husbandry and diet have been demonstrated to play a part. There is a seasonal variation, most blood spots occurring in the Spring. Experiments relating the occurrence of blood spots with age provide conflicting results. In a majority of cases blood spots decrease with age but a significant minority show an increase. Heavy breeds are more likely to produce blood spots. Certainly when judging contents classes I have noted many spots in the dark brown eggs.

Bantams usually exhibit fewer blood spots than the large fowl and they are a rare occurrence in waterfowl eggs. These observations may, however, be due to the fact that it is more difficult to pick out blood spots in the dark shelled eggs if they are candled prior to the show. An accepted figure for the occurrence of blood spots is 2-4% of chicken eggs when selected at random.

The Albumen

The 'good' white is even more difficult to quantify than the yolk. Its goodness lies in the 'solids' of the colloidal solution. As in the yolk the composition of the albumen varies in the different types of poultry.

Table 12 Shows the percentages of the chemical constituents of Albumen

Constituent	Poultry Type			
	Chicken	**Turkey**	**Duck**	**Goose**
Water	89.9%	86.5%	86.8%	86.7%
Protein	10.8%	11.5%	11.3%	11.3%
Lipids (fats)	0.03%	0.03%	0.08%	0.04%
Carbohydrates	0.9%	1.3%	1.0%	1.2%
Inorganic matter	0.6%	0.7%	0.8%	0.8%

Compared to the yolk there is more water but less lipids. There can be considerable variation in solid content between individuals of the same species. This suggests some degree of genetic control especially as individuals tend to produce whites of a constant composition. This probably lies in the number and effectiveness of the glandular cells in the oviduct wall. As indicated in Chapter 2, the four layers of the

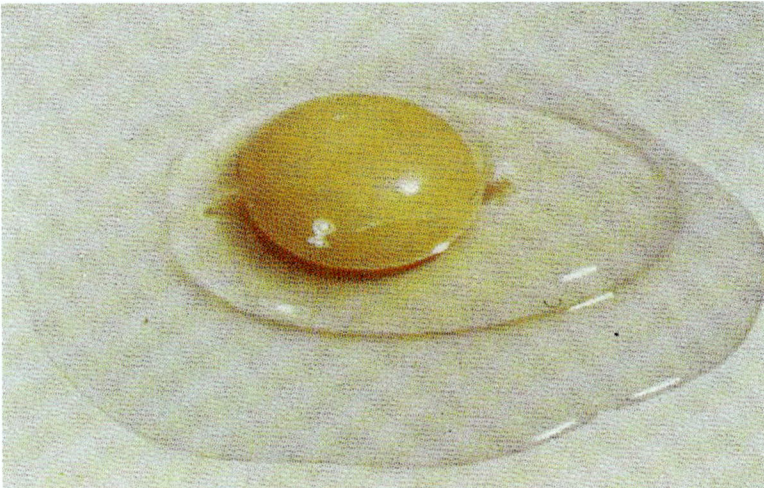

Well defined albumen

albumen – outer liquid, middle dense, inner liquid and chalaziferous – do differ slightly in the amount of water present. The chemicals found in greatest concentration in the albumen are sulphur, sodium, potassium and chlorine. Also present is lysozyme, which is the eggs natural defence against invading microbes.

All this makes the albumen good to eat but it is a quality that cannot be assessed with the naked eye. Candling will show evidence of 'meat spots' and provide a guide to freshness from the size of the air sac. The broken egg will further confirm freshness. The albumen in a stale egg is runny and fails to support the yolk The showman gives some consideration to the size, attachment and evenness the chalazae. These, however, do not impress the consumers. They certainly prefer a clear white, as near to a waterfowl albumen as possible! Unfortunately some poultry feeds do cause a slight green tinge in the albumen which is undesirable. Brown egg layers seem particularly prone to this fault. A yolk well positioned in the centre of the thick albumen is ascetically pleasing and looks good in the frying pan. If this is surrounded by a spread of thin albumen – not too much then the picture is complete. Certainly the clear definition of the thick and thin albumen is a judging point on the show bench. Because of the different composition of waterfowl albumen it usually appears clearer and better defined than that of chickens. This gives Waterfowl eggs an advantage in open competition. It is for this reason that the Poultry Club Show rules state that 'waterfowl contents shall not be judged in the same class as hen contents'.

Chapter 12

FRESH OR STALE

One challenge showing eggs has for me over the showing of birds is that every show requires a fresh selection of eggs. Hens come in and out of lay, so the winning eggs of last months are not available.

A change in the weather interrupts the laying sequence of a bird, so the final egg for a plate of three or six fails to appear or is mismatched; Anxious layers barge into the next boxes, scratching the surface of the previously laid egg.

Before each show a new clutch of eggs requires sorting and matching into the combination required to for the show schedule. Schedules can differ considerably from show to show. In contrast, a bird's class invariably remains the same and a good specimen can be shown over a considerable period of time with more or less consistent results.

Contents Deterioration

I do not say that I never show eggs twice. In the summer, if two shows are close together, such as, the Bath and West and the Surrey County shows, I will use the same eggs. In winter, the same eggs could be shown over a two-week period as not too much deterioration takes place. Deterioration is due mainly to evaporation. The yolk and albumen together contain about 65% water and when this evaporates through the porous shell and membranes from the air sac content changes occur which can be identified with the process of ageing.

Fresh Egg

Small air sac, firm
white, domed yolk

Shell bright colour
clear cuticle

Stale Egg

Large air sac, runny
white, flat yolk

Shell dull colour
translucent cuticle

Prevention of Contents Deterioration

The rate of evaporation is affected by the ratio of the area of shell surface and the volume of the egg. Certain egg shapes lose water quicker than others. It is also affected by external conditions. For example, eggs stored in a warm, dry environment will age faster than those kept in a cool, humid one. It is well known that eggs will keep longer in a refrigerator and if this also contains a dish of water little evaporation will occur at all.

Commercially eggs are stored at 0°C to –1.5°C with 80% – 90% humidity. A lower temperature would freeze the contents and the ice crystals would destroy the egg membranes. Surrounding eggs in dry sand or sawdust can help reduce evaporation.

Traditionally eggs were preserved for cooking by immersing them in either lime water or water glass (sodium silicate). The former caused a protective layer of calcium carbonate to form on the shell, both discouraged gaseous exchange and the activity of micro organisms. It has to be said that such preserved eggs like any others kept for some time do lose their taste. This is due to chemical changes in the lipid (fat) content of the yolk.

Changes Associated with Ageing

As evaporation proceeds, the air sac increases in size and the egg contents decrease in volume. These changes cause the enclosing membranes to stretch in some places and relax in others, thereby allowing the contents to 'flop' within the shell when the egg is shaken sharply. This becomes more evident by the white being less firm as certain proteins within the albumen denature; a process related to changes in the pH (acidity) brought about by the release of carbon dioxide.

During development the egg produces carbon dioxide as a by-product of metabolic processes and when laid it contains more of the gas than the surrounding atmosphere. Its presence in the egg produces a pH of 7.6 but this increases to 9.7 in a matter of days as the carbon dioxide leaves the egg. The rate of loss is most rapid if the egg is kept in warm conditions. Such high alkaline conditions prompt the breakdown of the mucin fibres in the albumen and so it loses structure and appears runny when the egg is cracked open on the plate. One of the tests for freshness involves measuring the area covered by the broken egg.

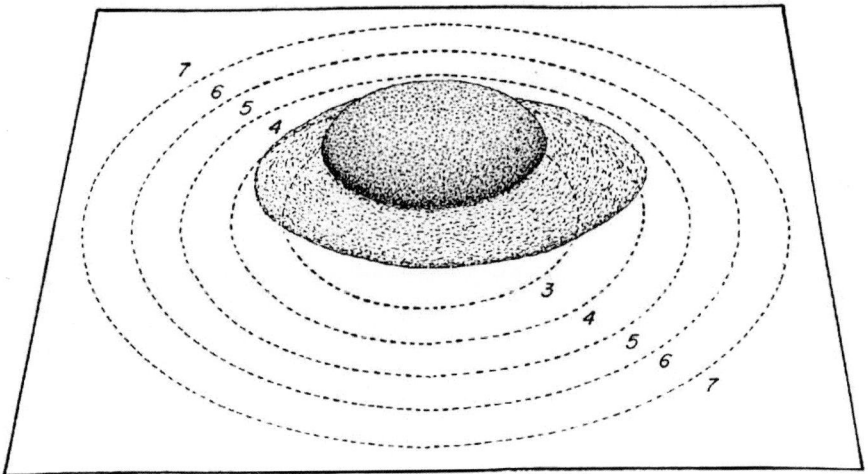

Fig 24 Apparatus for Measuring the Area of Albumen

As the albumen loses structure the thick and thin areas lack definition and they are less able to support the yolk. The height of dense white is thus another indicator of freshness and is measured by the Haugh apparatus.

Fig 25 The Apparatus for Measuring the Height of the Albumen

The ratio of the height and diameter of the albumen gives a figure called the **albumen index.**

$$\frac{\textbf{Height of dense albumen}}{\textbf{Average diameter of albumen}} = \textbf{albumen index}$$

The albumen index is a figure that can be used to quantify changes over time. The Haugh apparatus can also be used to measure the height of the yolk as another indicator of freshness. As the albumen loses structure there is less support for the yolk and it sinks.

The wrinkling of the egg membrane that surrounds the yolk is another indicator of age. At the time of laying the yolk takes water from the albumen by osmosis – a process that results in balancing the water content of the two areas. As water enters the yolk it increases in size so stretching the vitelline membrane. Once evaporation of water from the albumen via the shell starts the albumen has less water content than the yolk so water flows from the yolk decreasing its size which results in the stretched egg membrane becoming wrinkled.

Estimating the Freshness of Contents from the Outside

All the foregoing tests for freshness require the egg to be broken open, but the changes described can be inferred by handling the whole egg. To determine a measure of liquidity judges jerk an egg past their ear, listening for a tell tale 'flip'. If shaken too vigorously, however, even a fresh egg's contents can move within the membranes. At shows where no barriers are placed between the exhibits and the public, winning eggs can be shaken by all who pass with the result that by the end of the day they too will 'rattle' and the judge's reputation held in dispute. Fresh eggs, when sent to a show by rail or parcel post, also tend to end up rattling on the show bench.

I have often been surprised when judging, to find a fresh-looking egg has rattled – 'fresh-looking' might sound subjective, but to the practised eye there are considerable differences between fresh and stale eggs.

Brown eggs take on a dull almost greenish hue when exposed to bright light for any length of time. White eggs quite often develop translucent patches in their shell with age. All eggs lose their bloom, becoming dull and lifeless with age.

When faced with a stale-looking egg, I break it open, as empowered to do by Poultry Club rules. A large air sac, runny white and a flat yolk with wrinkled membrane would then confirm my suspicions. However, the 'jerk' test is not infallible. Some eggs, especially those with a broad girth, laid by large fowl, have poorly attached membranes so rattle the day after laying.

Is there an alternative to the 'jerk' test? Assuming evaporation has taken place and that the now enlarged air sac is at the broad end, the egg will float broad end up in a pan of water. This will not indicate, however, the egg that has been stored in a refrigerator if other signs of ageing are absent. I have known instances of eggs being shown and winning after six months of cold storage! However, care must be taken not to heat up such eggs too quickly as this affects the internal volume and hence they will 'flop' when jerked.

If ageing has been coupled with infection by micro-organisms through the shell then tell tale odours and discoloration of the yolk and albumen will be evident. The egg is then well and truly addled.

WHERE ARE YOU GOING TO MY PRETTY MAID? TO LAY EGGS FOR THE WOUNDED YOUNG SIRS SHE SAID.

The National Egg Collection,
154, Fleet Street, London, E.C.

SECTION 3

SHOWING EGGS

Chapter 13

A BRIEF HISTORY OF EGG SHOWING

Early Shows

At most poultry shows now there are classes for eggs which are judged according to the Poultry Club standard. This standard has been developed over the years, the details of which have their origins in the egg laying tests of the 1920's and 1930's. Groups of hens would be kept at a farm where the number and sometimes the quality of their eggs were recorded. Trials could last for six months or more. The 'quality points' looked for were shape, size, shell texture and colour, exactly the points required by the present day standard. These trials were held as a competition between birds but it was not too long before competitions were held just for the eggs.

The early shows were the show piece for competing 'commercial farmers'. Fig 26 shows an egg show from the early 1930's in America. Classes were for packs of eggs, often a dozen or more. Here contents were included as well, the judge having selected three eggs at random from the pack of fifteen with which to judge the contents. The equivalent class on the present day show bench is the external/internal class but this only involves one egg.

In the post war period competitions involved trays of 30 eggs and even boxes of 30 dozen, the standard box size sent to the egg packing stations. Competitions between commercial producers still exist today. In November, W.R. Curnick Ltd. of Burbage, Wilts won the best tray of eggs at the annual Wessex N.F.U. poultry conference. 'Poultry World' regularly reports upon awards won by egg producers.

Fig 26 A Pre-war American Show Bench

Rules for Judging

Rules for judging such competitions were developed. By 1946 the United States had clear descriptions for egg quality which graded eggs AA, A,

B, or C. Descriptions covered the shell, air sac, albumen and yolk. In Britain, the Ministry of Agriculture Fisheries and Food published a booklet entitled, 'Testing of Eggs for Quality', which described the necessary criteria for first and second quality eggs.

Such rules, however, were not in place at the Agricultural Shows around the country. Mr J. Gatecliff wrote the following letter to a poultry magazine. In it he called for a 'National Egg Club' to be formed.

'It is about time an egg club was formed for those of us who specialise in eggs –undoubtedly thousands of others would welcome and endorse this clarion call.

Not only the public, but most exhibitors require education upon the importance of the beautiful egg and the definite conditions under which they sho uld be shown. Unfortunately many show committees either consider egg exhibits of very secondary consequence, or are flagrantly ignorant that as much skill is required in producing the prize egg as the prize bird.

Too frequently at shows the egg displays are placed in some out-of-the-way corner, the top of rabbit pens, or else to high for people to see. Occasionally they are staged in box lids, baskets or plates of various sizes and decorations, etc. Such staging is an insult to the exhibitor who has taken infinite pains over his eggs. Another and the greatest indictment against many show committees is their slipshod manner of scheduling egg classes, giving either insufficient or no conditions of entry, or, worse still, misleading definitions.

Looking over the eggs at a South-Eastern county show, I pointed out to the egg steward an egg in the first award that appeared of very doubtful age. He replied, "Well! What does that matter? The schedule gives no age limit!" At a North county agricultural show the qualification was **Fresh** and the eggs were to be judged irrespective of size, symmetry, quality and soundness.. Here apparently first-grade 2 oz eggs were not necessarily the standard. At a recent congress for chefs' a fresh egg was considered as such for 21 days.

A "National Egg Club" would rectify all these anomalies, and we should have standard and comprehensive rules to guide us.

Furthermore is a competent judge of birds an equally competent judge of eggs? The judging of eggs is a distinct and separate art.

A gentleman told me only a few weeks ago that he had been a judge of eggs (even including some of the big shows) for over twenty years and had never broken an egg at any one of them!

Such feelings were obviously shared by others for rules for the showing and judging of eggs were put in place, but not by a National Egg Club.

The Poultry Club Year Book of 1948 published instructions for judging eggs at shows held under their rules. It also included a scale of points for judging the external egg and internal egg. This standard remained until modified for the fourth edition of the Poultry Club Standards in 1982.

Egg Judging Standard (1948)

The Poultry Cub has accepted the following Standard and Scale of Points:

External

Shape	20
Size	20
Shell texture	20
Colour	20
Freshness, bloom and appearance	20
	100*

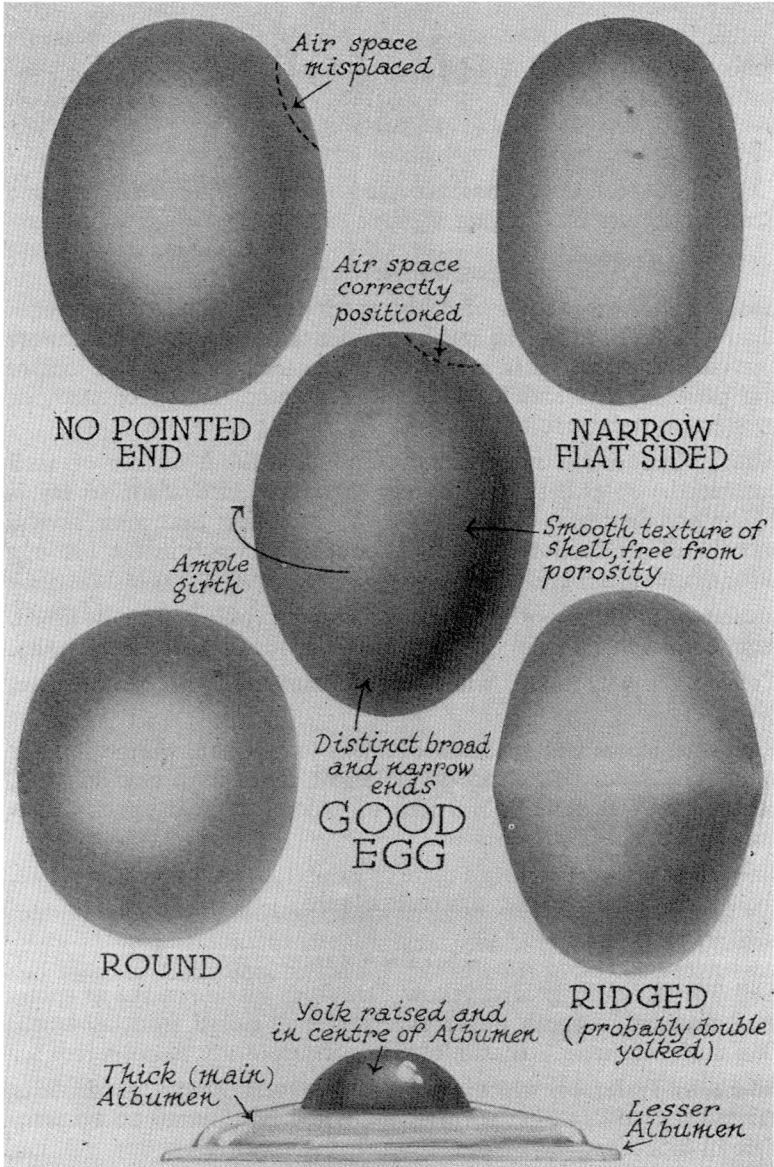

Illustration from 3rd edition of Poultry Standards, 1960

The above may be a maximum for each egg, or for a plate of eggs, whatever the number; or for a plate of eggs add 5 points more for each egg for matching and uniformity.

Internal

Yolk	40
Albumen	30
Chalazae	10
Airspace	10
Freshness	10
	100

The standard was based upon that used by W. Powell Owen, president of the Poultry Club, 1939-52. Writing in his book, *The Complete Poultry Book*, he states:

> **'Many exhibitors do not show live poultry but specialise in breeding for ideal eggs which they exhibit. Small poultry keepers also like to support egg sections at agricultural events. Classes are for single eggs, threes, sixes, 'sittings' (12) and commercial displays.'**

During Powell Owen's involvement with the Poultry Club and the Domestic Poultry Keepers' Council egg shows developed apace. He admits to regularly judging shows of 1,000 – 1,750 entries. His highest was 2,165! Individual classes often had 150 entries. His record contents class was over 200. H. Murche, writing in the 1949 *Poultry Club Year Book*, writes of the boost to egg showing that resulted from the restrictions on livestock showing due to fowl pest; he quotes entries of 600 at Hammersmith, 450 at Willesden, 300 at Orpington. Where now the London shows?

Egg Showing 1960 to 1980

Thereafter the great divide between commercial poultry keeping for egg production and the small producer occurred. The latter slowly disappeared. Some of the pure-bred stock was kept by back yarders and fanciers and the modern show scene became established. Showing birds for standard perfection took precedence over utility and by the 1960's egg sections at shows were very poorly supported or indeed reorganised. Pleasingly, today the situation is being redressed, due to the efforts of the small band of 'egg men'. One, Max Butler, did more than anyone during the late 70's and 80's to re-establish egg shows.

It was he who wrote at length at the time in various magazines about eggs. As a member of the Poultry Club Council he took it upon himself to promote the egg section at Shows and encourage the training and use of specialist egg judges. I was one such trainee. He also was responsible for the 1982 modification of the standard which involved a change of emphasis in the external points for shape and size and the internal points for yolk and albumen. In the former he regarded shape (type) as paramount and so it was allocated 25 pts, size was thus reduced to 15 pts. However, size was defined more closely and a maximum of 1½ ozs (42.5 g) set down for bantam eggs. In the internals yolk and albumen were given equal status, 35 pts each.

Egg Judging Standard (1982)

External

Shape	25
Size	15
Shell texture	20
Colour	20
Freshness, bloom and appearance	20
	100*

Internal

Yolk	35
Albumen	35
Chalazae	10
Airspace	10
Freshness	10
	100

Egg Showing at the end of the Century

By the end of the 1980's three 'egg' only shows had become part of the poultry calendar. Quite appropriately they all were held in Spring – the Colchester egg show, the Preston egg show and the Wessex egg show. All boasted entries of well over 500 plates catering for breed clubs, plates of 12, 6 and 3, as well as singles, external and internal. Run by egg enthusiasts Len Frost and E Dellow in the East, Bill Oldcorn and Paul Kerfoot in the North and J Breslain and M Thompson in the South they provided an attractive spectacle and keen competition. At the Wessex there is a team competition in addition to the individual competition. Competitors winning prize cards score points for their Club, be it Ashdown Forest, Reading Hants and Berks, Frome or Oxford. All over in a matter of three or four they enable the egg enthusiasts to talk in peace and quiet without constant interruption from extrovert cockerels. It was at a Wessex egg show that the idea for this book was converted into a reality by Dr Joseph Batty.

Since the Jubilee in 1997, the Poultry Club has supported Regional and Championship Shows at which section winners win embossed cards – five of which can be exchanged for a Trophy. Eggs have been one of these sections and this together with the vouchers on offer at other shows have encouraged egg showing. In 1999 George Taylor from Cumbria won 16 bronze awards, quite an achievement. With the publication of the 5th edition of the Poultry Standards in 1997 some further changes were made, plus considerable amplification. Exactly what is required under each heading was explained in detail. There was also recognition of eggs other than those of hens and bantams. The only points change concerned the internals. Yolk and albumen were reduced to 30 pts each and freshness increased to 20 pts. The extra points, refer, however, to the freshness of the yolk and albumen so it would be fairer to say the points had been reallocated under a different heading. The required standard was, in essence, unchanged.

Developments in the 21st Century

As the popularity of keeping a few garden poultry increased so, too, did the popularity of showing eggs. Many new fanciers begin by showing eggs at their local Agricultural Shows. These shows had a history of staging egg classes, often in the domestic produce classes. They also staged painted, displayed and decorated eggs. These classes began to be found on a regular basis at poultry shows and the Poultry Club produced a standard for each of the classes. These were published in the sixth edition of poultry standards in 2008.

More stand alone egg shows have been started by poultry clubs. As well as, classes for turkey eggs, quail eggs and the most unusual shaped egg, new combination classes have been included in schedules. For example, pairs of eggs, large and bantam of same colour or different colour. In addition to three different coloured eggs a class for three eggs to include a waterfowl, large fowl and a bantam. Classes for true bantam eggs became quite common.

The new egg shows each developed their own flavour. The Craven Poultry Club introduced the 'Guineas' Show at which substantial prize money was paid in guineas to winners. The Welsummer Club organised a National Welsummer Egg Club Show. A very recent innovation by the Poultry Club has been the staging of a National Egg Show at the same time as the National Poultry Show.

In 2012 the Poultry Club agreed to sponsor 'stand alone' Regional and Championship Egg Shows. To qualify the former require 200 plus entries and the latter 400 entries. Shows that affiliate can award a Gold award for best eggs, a Silver card for the reserve and Bronze cards for: plate of large fowl eggs; single large fowl egg; plate of bantam eggs; single bantam egg; plate of waterfowl; single waterfowl egg; plate of turkey eggs; single turkey egg; juvenile exhibit; contents; decorated or painted egg; The last two bronze awards cannot compete for the Silver and Gold awards.

List of Egg only shows in 2016

Royal Championship	Championship	Regional
Royal Cornwall	Summer Wine Show	South Devon Fanciers
Royal Bath & West	Preston & District	Duke of York
Royal Cheshire	Southport & Ormskirk	Kent Fanciers
	Wessex Egg Show	Stalybridge
	Craven Fanciers Guineas	Bolton & District
	Scottish National	Peebles
		Bugle
		High Peak
		Norfolk Poultry Club

All this is a far cry from the 1970's when, if egg classes were staged, the classification was limited to about six classes judged by the poultry judge who was first to finish. The number of qualified egg judges now must be at an all time high. Max Butler would be very proud of what has been achieved during the last forty years.

Poultry Club Bronze Award Card

THE POULTRY CLUB
OF
GREAT BRITAIN

Patron: H.M. QUEEN ELIZABETH, THE QUEEN MOTHER

THE 'POULTRY WORLD' DIPLOMA

BRONZE STAR

CHAMPION EGG EXHIBIT

REGIONAL CHAMPIONSHIP SHOW

Poultry Club Voucher Card

THE POULTRY CLUB

OF GREAT BRITAIN

Patron: H.M. QUEEN ELIZABETH, THE QUEEN MOTHER

At .... READING .... Show

held on the ...17th... day of ...AUGUST... 19..85

THIS CUP VOUCHER
AND CERTIFICATE OF MERIT

was awarded to

EXHIBITOR W G V THOMSON CLASS 43 PEN 137.

FOR BEST EXHIBIT OF EGGS IN SHOW

Show Secretary   J. Breslain _Judge_

SEE BACK OF CARD FOR INSTRUCTIONS

Chapter 14

SHOW RULES

The Exhibiting of Eggs

Instructions to Show Executives and Panel Judges issued by THE POULTRY CLUB.

These instructions concern the Staging and Judging of Egg Classes.

The PCGB standard descriptions and Scale of Points are included in *British Poultry Standards.*

(1)   Each egg exhibit should be staged on uniform plates provided by the Show Committee. A little softwood sawdust or similar material on the plate helps in arranging the exhibits to the best advantage. Plates should not be fixed tp the table to allow judges to move exhibits to the front of the show bench for ease of inspection when judging.

(2)   The show should appoint a steward for the Egg Section whose name should appear in the schedule.

(3)   Where possible eggs should be exhibited out of reach of the public. Exhibits should only be handled by the judge, steward or exhibitor.

(4)   At shows where duplication of entries is permitted, this applies only to duplication into Challenge classes.

(5)   The colour classification for large fowl or bantam egg classes that can be shown are as follows:
Dark Brown; Light Brown; White; Cream; Blue. Any other colour (A.O.C.) should be used for other colours such as green and khaki etc (N.B. *'Tinted'* is not recognised by the PCGB and the Light Brown class is appropriate)

(6) Shows have the discretion to offer any combination of eggs in each class. Plates of 1, 3 or 6 eggs are the usual combinations.

(7) Show Committee should state Large fowl or Bantam in the schedule for egg exhibits and include the wording; 'Bantam eggs not to exceed 1.5 ounces or 42.5 grams'.

(8) Separate classes must be given to waterfowl, turkeys and guinea fowl etc. and for any classes that a Breed Club may deem necessary.

(9) Judges should not reclassify egg exhibits to classes for the correct colour prior to judging. It should be stated in Schedule that judges are not able to reclassify eggs.

(10) Only fresh eggs should be entered in the egg classes, and Judges should disqualify any exhibit which, in their opinion, does not conform to this ruling. Judges may, at their discretion, break open an egg to test that it is fresh.

(11) Mottled eggs should be judged according to their ground colouring, unless the schedule states otherwise.

(12) *'Distinct colour'* classes (separate colours e.g. white, brown and blue) should include eggs from the classification as shown in rules 5 and 6. Classes for *'Different colour'* classes can be for shades of similar colours e.g. plates of eggs where each colour is brown.

(13) Cream eggs should not be judged in classes for 'White Eggs' unless specially stated in Schedule, i.e. 'White or Cream'.

(14) Eggs may be washed but not be artificially polished or over-prepared by heavy rubbing. No artificial colouring is permitted.

(15) If appropriate, shows should include separate contents classes for Large Fowl, Bantam and Waterfowl eggs.

(16) When judging contents, it is important that the spread of the egg is not restricted. Therefore, it is recommended that china or hard plastic plates are used for this section.

(17) Shows may offer classes for External/Internal eggs. Both elements should be judged separately according to the PCGB Standard.

(18) Classes can be offered for decorated eggs, painted eggs and displayed eggs. Only in classes for displayed eggs are judging points awarded for the quality and matching of the eggs which therefore should not be hidden by the display.

(19) The maximum size of a displayed exhibit should be stated in the schedule.

The main changes in rules from those in force in when the first edition was published are the non-recognition of tinted as a colour and the forbidding of judges to reclassify eggs before judging.

In the first set of rules set out by the Poultry Club Administration Board in 1948 the following was also included:

"Executives should arrange for broken eggs to be given to Hospitals arranging for suitable containers for their transport."

How times have changed! After the Salmonella scare of the '80's no egg steward would dare to make such a suggestion.

Chapter 15

Show Classification

The present day egg show has a classification similar to that outlined by Powell Owen in the 40's. However, plates of 12 are only staged at a few shows; ones and threes are the usual classification, although some shows in the North have classes for four eggs. Colours are shown separately, my ideal is as follows: Dark Brown, Light Brown, Cream or White and A.O.C. This latter class would include blue green Araucana eggs, plum and any other 'off colour'. Notice I make no reference to tinted, despite the fact that it is an egg colour in the poultry standards book. The reason for this is that tinted is an imprecise term. Tinted can be any shade from 'light brown' to dark 'cream'! Some even use the term to include any 'light' shade of colour. See page 65. If the term tinted has to be used then let the class be called tinted or cream and let white stand alone.

If these classes were denoted in the schedule as a single egg or plate of three eggs, then exhibitors would be justified in showing hen, bantam and indeed waterfowl eggs all together. This makes judging rather difficult for, although at present, all have the same standard there are characteristic differences. It is fairer to the exhibitors to have separate hen, bantam and waterfowl classes. Bantam eggs now must not weigh more than 1½ ozs (42.5 g) otherwise they will be passed. At some shows classes are scheduled for miniature fowl separately from true bantam eggs whose weight is under 1 ozs.

Egg contents classes are now a regular feature at shows. This was a class devised by Powell Owen to check, he said, 'the poultry keeper's feeding system'. The contents can be inspected for freshness or double yolk in the external classes if the judge so wishes but this is rarely done.

Present Poultry Club rules prevent the showing in the same class of chicken and waterfowl eggs because they are quite different in nature.

A class that is of fairly recent origin is the external/internal class. Here a single egg is judged first for external features and then for its content, the final placing being determined by the combined points score.

Powell Owen also introduced a class for three eggs of different colour, a challenge for the exhibitor to find three chickens each laying ideal eggs that match in size and shape. Some show schedules list 'distinct' colours which implies that the plate can not include shades of the same colour.

At the Bath and West Show a class is scheduled for a plate of three eggs, one waterfowl, one hen and one bantam. The aim is to have three eggs of ideal shape and in proportion despite their different size. There is no requirement for the colour to be the same.

A new class, but will it catch the imagination of egg exhibitors? Is it taking the show bench too far away from utility? Should we be providing classes for the ideal hatching eggs and for contents that suit the market demands rather than the more exotic combinations?

Suggested classification

Large Fowl and Bantams

6 A/C

3 Dark Brown incl. Marans/Welsummer

3 Light Brown

3 White/Cream

3 A.O.C.

3 Distinct contrasting colours

1 Brown

1 White/Cream

1 A.O.C.

1 Contents

Waterfowl and Turkeys
Plate of 3
Single
Contents

A colour classification for egg classes at a show

Dark Brown

Cream

Light Brown

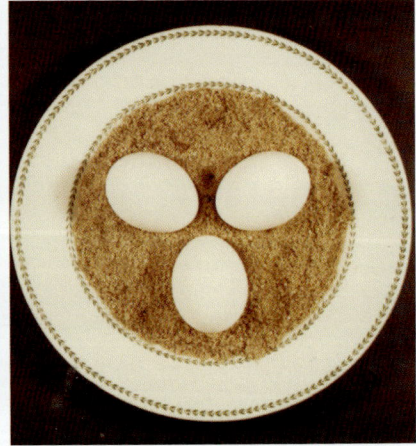

White

Chapter 16

ORGANISING THE EGG SECTION

Staging

Wen putting up the Poultry Show it must always be remembered that there are many poultry keepers who do not show their birds but specialise in exhibiting their eggs. It is therefore important that the exhibitors are properly catered for and that every attention is given to the egg section.

In addition to these fanciers, there is the general public who never fail to admire and show interest in this part of the show, and therefore it should be presented as attractively as possible. This means using a covering for the table tops and skirting round the sides. It is also important to site the section in a good light – so necessary for judging – and to check that the tables are level.

Tables covered in green matting or cloth set the exhibits off to best effect. Even white lining paper is better than bare trestle table tops. Depth of the table is important. Anything over three feet makes it difficult for the judge and visitor alike; Eggs at the back are not seen to good effect and those in the front can become dislodged when the judge examines those at the back.

To help the visitor, each class should be clearly marked with details of the colour and type of egg required. Different classes should be separated by tape or thin strips: Nothing is more confusing than being faced with a crowded, jumbled assortment of egg exhibits, not even experts can always distinguish Marans from Welsummners and tinted from cream without such guidance.

Space between individual exhibits is important. Overcrowding makes it difficult to appreciate the individual exhibit and certainly makes the judge's task harder. There should always be some 'free' space in which the judge can place his cloth and scales while examining the class.

Once prize cards are awarded the problem of over-crowding is exaggerated. It is often difficult to see which card refers to which exhibit, indeed at some shows the prize cards are so big the exhibits disappear beneath a kaleidoscope of colour.

Displaying the Exhibits

Plates should be used for the exhibits using a small amount of softwood sawdust on each plate to hold the exhibit in position. Polystyrene plates are best. Cardboard plates are adequate for entries of three and single eggs but the individual eggs should be on smaller plates. Many judges like to rearrange the exhibits and single eggs roll off large plates. Similarly, cardboard plates of 12 and 6 are not easy to move as they give

Table arrangement

way under the weight of the eggs. Some organisers secure the cardboard plates to the tables but this prevents judges from moving the plates and the drawing pin head often prevents a symmetrical arrangement of eggs on the plate.

Specialist classes require a little extra attention. Contents should always be broken on to 6ins china plates, though non-absorbent poly-styrene ones will not do as a substitute at one day shows. Saucers are not suitable as they prevent the white spreading to its full extent and the rim makes judging the height of the yolk difficult. Paper plates are a disaster, as when moved, the egg is likely to slide off and after a few hours so much water has been absorbed that the yolk looks like a wizened apricot.

Painted or decorated eggs are shown off to best effect if supported in the vertical position by a cardboard ring or stand. Attractively displayed eggs must have space – too often, they resemble a colourful jumbled heap at the end of the egg tables.

Information for the Public

The egg section creates a lot of interest from the general public. Whilst judging one can be almost overwhelmed by their enquiries. A couple of large cards outlining what the judge is looking for in the external and

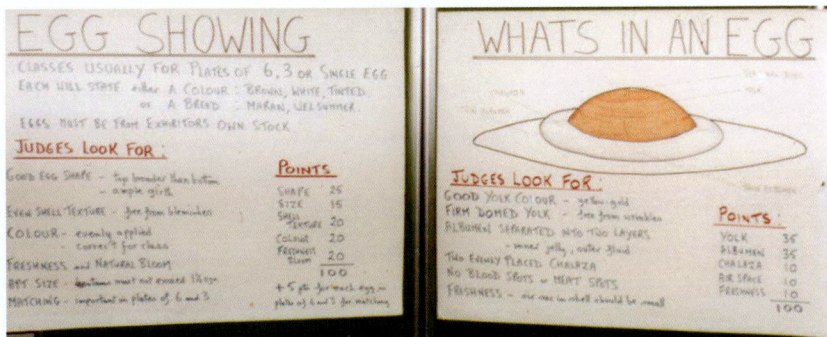

Display boards

contents classes can be helpful. They don't take a moment to prepare and answer the visitor when judging is completed.

While making things attractive for the public one must guard against theft or damage. I have seen tables covered by transparent poly-thene that screens the eggs but detracts from the exhibits and, at summer shows, produces condensation. A roped off area at arm's length around the tables is the best solution. A notice 'Please do not touch' helps to keep straying hands off the exhibits.

The Egg Steward

A senior steward should be appointed to be in charge of the section, who will be responsible for putting up the display, checking the plating of the exhibits the show and clearing up afterwards. Some one should be keeping an eye on the section at all times as egg exhibits are vulnerable to damage and theft.

At some shows, with the owners permission, the steward sells the eggs at the end of the show. It is therefore important to see that the entry forms are clearly marked in the appropriate place 'Eggs to be Returned' or 'Eggs Donated to Show Funds'. It is also helpful is exhibitors state the name of the breeds whose eggs they are exhibiting in order that display cards can be affixed to the exhibits if required.

The Judge

This is the individual who can make or mar the section, so employ the best. A specialist should be the order of the day. No longer is it acceptable for the soft or hard feather judge to adjudicate in the egg section unless he is qualified for that purpose. There are many egg judges now with Poultry Club qualifications so be sure to employ one. Remember also that the selection of the judge can have an influence on entries, therefore book early and get the judge you want.

Chapter 17

EGG SHOWING GUIDE FOR EXHIBITORS

What are the best eggs to show? The answer is, any eggs which measure up somewhere near to the requirements of the standard for eggs, no matter what the colour or from which breed they come. All have an equal chance, not only of winning a class, but of being considered for the Best in Show award.

How then does one come by the desired type suitable for the show bench? Chickens lay various shaped eggs, all too few suitable for showing I'm afraid, but obviously the more birds you have the better selection you have to choose from. On the other hand it just so happens that even amongst the smallest of flocks there could be a layer that will produce a suitable egg.

It is thought by some that you have to keep pure breeds to be successful at showing. This is not necessarily so, as people who keep a few hybrids just for eggs for the table have won many a major award at the egg shows. All they may miss out on are some of the colour classes such as the very dark brown, blue and the distinct colour classes etc.

The real answer to obtaining eggs suitable for showing is to breed for this purpose. When one is breeding for reproducing the species to the requirements of the breed standard, one has to give much thought to the make up of the breeding pen, selecting only stock suitable for this purpose. Having done this, one has to accept the eggs laid by the birds in this pen, whether they are good, bad or indifferent, if it is show progeny that are required.

When breeding for the purpose of producing ideal eggs, again it is attention to the breeding pen that is all important, but on this occasion one doesn't have to worry so much about the show points of the birds but only about the eggs that they lay. It therefore follows that the eggs you select for incubation should be only those that you would put on the show bench. This, of course, is no guarantee that you are going to reproduce only birds that lay suitable eggs for showing, but it does help, and is a step in the right direction.

Obviously, parent breeding stock of known background is essential in any breeding programme and breeding for eggs is no exception. Strain, too, is so important.

Classification of Breeds

In this country we have two classifications for large fowl, namely, Light Breeds and Heavy Breeds The Light Breeds are comprised of mainly

Marans
heavy breed

Araucana
light breed

Ancona
light breed

Welsummer
light breed

Mediterranean breeds. These are generally considered non sitters and lay white eggs. The Heavy Breeds are comprised of mainly Asiatic and American breeds, they are considered sitters and lay brown or tinted eggs. Dorkings, Sussex and Indian Game however, are of British origin.

Some countries have three classifications – Light, Medium and Heavy, and therefore there are certain exceptions. For example, the Welsummer is this country is listed as a Light Breed and lays a brown egg. In Holland, its country of origin, it comes under the heading of Medium Breed.

Most breeds have a miniature counterpart for which classes are staged. The requirement for these classes is that the eggs should not weigh over 1½ oz/42.5 g. There is a group of fowl that do not have any large fowl counterpart – True Bantams. They are small birds and so lay a tiny egg. Some specialist show put on separate classes for eggs under 1 oz.

Breeds and their Egg Colour

Although there is a wide and varied selection of breeds from which to choose, it is probably true to say that the majority of eggs seen on the show bench today come from the following pure breeds and their crosses:

Dark Brown:	Barnevelder, Marans, Welsummer
Brown:	Croad Langshan, Orpington, Rhode Island Red
Blue, Green & Olive:	Araucana, Cream Legbar
Tinted:	Australorp, Indian Game, Light Sussex, North Holland Blue, Plymouth Rock, OEG, Silkie, Wyandotte
White or Cream:	Ancona, Andalusian, Campine, Hamburgh, Minorca, Leghorn, Spanish

exactly what it says. An exhibit of two shades of brown together with a white or cream egg is not what is required. Poultry Rules now allow for the showing of shades of the same colour under the class heading – **three different colours**.

The obvious answer to that third distinct colour is the blue or green egg of the Araucana. This is a breed that all egg showing enthusiasts should keep and breed. It is also an excellent layer. With a pen of Araucana together with the layers of brown and white eggs, the collection of three distinct colour eggs presents no problem. The challenge is, of course, matching size, shape and texture from three different breeds.

The egg of the Araucana is undoubtedly one of the most attractive exhibits to be seen on the show bench, and indeed on the breakfast table. It is worthy of note that unlike some breeds of fowl that lay eggs where the colour is only pigmentation on the surface of the shell, the colour of the shell of the Araucana egg is the same throughout.

Contents Classes

These are always interesting classes to show in and they are usually well supported. Here the exhibitors are unable to see their entries before judging which means a little homework is necessary prior to the show if the competitor is to stand any chance of success. Obviously, it is advisable to know the make up of an egg.

Having made an entry, one should start recording the daily collections of eggs. By opening one or two eggs from different birds a contents should be found somewhere near to specification. It will be from these birds that you should select your exhibits. In this way you will increase the chances of having a good show contents on the day. A fresh egg should used but not one laid the night before or on the day of the Show. A very fresh egg when broken onto the plate does not show the yolk and albumen off to advantage as the albumen lacks viscosity.

Egg contents

How the birds are fed does, of course, influence the contents of any egg both in taste and appearance. One should always aim for a balanced diet, which means not too much of any one ingredient. While the layers love a cabbage or the like hung up in the run which is ideal for keeping them occupied, from a show content point of view, too much green food taken in excess could make for a greenish tinge in the albumen. Likewise too much protein could make for poor shells. Lack of shell forming materials will not help to make sound shells so necessary for good contents.

Finally, with the exception of the very dark brown eggs, one can always candle the exhibits as a final check before despatch to the show.

Show Preparation

With a show coming up, now is the time to check the nest boxes. Better still, give them a thorough clean out and replenish with new litter. A layer of softwood chips with a generous supply of soft straw makes ideal nesting material. Nothing is more annoying than to find soiled eggs when making the daily collections, either by dirt taken in by the feet of the bird or by droppings. It is important that you check for cleanliness at every visit to the nest boxes.

Washing eggs is permitted under Poultry Club show rules, but as eggshells are porous, avoid doing so if possible. Some shells do not always take kindly to it, particularly the dark brown variety where colour is pigmentation and could easily rub off. If you do have to wash, make sure in the drying process you do not polish them as this could lead to the judge passing them. Wiping carefully with a slightly damp cloth should be all that is necessary just to remove any dust and leave the egg with it's natural bloom and ready for the show bench.

Make collections at least twice a day putting aside any with show potential – there will not be many and some days none at all – and store in a cool place. The day before the show you can then sort out the 'probables' from the 'possibles', make up your entry even to the extent of plating them if necessary.

Remember that there are 5 points for each egg for matching and uniformity. It is so important therefore when showing a plate of three to see that all three eggs really do look alike. Ideally, a plate of eggs should all come from the same layer to obtain maximum uniformity. Though it should be said that uniformity from a bird that lays four eggs in four consecutive days is less likely than from a bird that lays every other day. Eggs laid by the more prolific layer could vary in colour, and the last egg of the sequence of four could be slightly smaller than the first.

In the dark brown egg breeds such as the Marans and the Welsummer, colour is at its best when the bird first comes in to lay as a pullet, or when she has recommenced to lay after the moult when she is known as a yearling. Generally speaking, the more a bird lays the lighter the brown egg becomes. Darkness in colour of the dark egg breeds is a matter of strain and there are some strains that retain the dark pigmentation longer than others.

Showing white eggs is a real challenge. A good plate of white shown to perfection is a joy to behold but it is one of the hardest exhibits to plate and one that needs the most care. Any mark or blemish stands out like a sore thumb, and here cleanliness in production is an absolute must. True, there are some white egg shells that stand up to washing better than others, but I would say more are spoilt by washing as it emphasises the porosity in the less sound shells and thereby rendering them useless for the show bench. Perhaps this is why one does not see so many classes for white or even plates of white in classes for white or cream.

For those who do have to wash white eggs but fail to get out an obstinate stain try lemon juice.

Do's and Don't's for Egg Exhibitors

DO

- Inspect the nest boxes daily for cleanliness
- Use soft litter for the nests so as not to mark the shells.
- Collect the eggs more than once a day.
- Use a proper egg box when collecting so as not to cause any damage.
- Watch your feeding prior to a show.
- Store in a cool place.

DO NOT

- Overfeed your birds as they could become fat and lay defective shells.
- Feed too much green food as this could make for the wrong colour contents for the show bench.
- Feed too much animal protein as this could make for thin shells.
- Wash eggs if you can avoid doing so, as this in some cases can have an adverse effect on the shell.
- Polish an egg. A judge could pass an exhibit if he thinks the shell has been polished
- Send the same eggs from show to show.

Chapter 18

JUDGING EGGS

So you want to be an egg judge? I can recommend it as a rewarding and exacting task. Judging eggs as a first steps in the poultry judging ladder is a sensible introduction since the standard and judging points are not extensive. One can marshal the pro's and con's privately in a glance without the publicity that judging birds brings; No need for assessment over rows of cages and of handling unwilling prima donnas whose characteristics seem to change with every different angle and pose.

First Steps

The first step is to steward for a couple of recognised judges, asking them to explain the reasons for their decisions. Their views and preferences are personal interpretations of the egg standard so next acquaint yourself with the official Poultry Club Standard. By careful reading it should become clear how objective your mentors were in their judgements.

It is now time to consolidate the theoretical knowledge so far gained with the practical business of showing. It is only by sorting and resorting a selection of eggs for showing in various classes that one begins to appreciate the variety of egg characteristics.

A plate of three may not seem difficult to attain but rarely do three eggs match, even closely, for shape, size, texture or colour. Once your choice is made and is on the show bench surrounded by everyone else's selection, it is time to take further stock of the eggs.

After judging is over, inspect all the eggs and compare your placings with the judge's. Try to work out why the cards have been awarded as they have. Avoid making snap judgements of your own and becoming involved in hasty criticism of the judge's choice. The longer you retain the ability to be receptive to other's interpretations, the better judge you will eventually become.

When your judgements coincide with those of the judge's in about 75% of the classes, it is time to consider accepting your first judging appointment. Select a small local show and try to enlist an experienced judge to steward for you. They will be on hand to put you right should the occasion prove too much. What follows should become a precisely followed routine. A standard step by step procedure is essential if decisions are going to be more than just the view of the moment. Above all things, a judge must be consistent in their awards.

Materials for Judging

Make sure that you arrive at the show with all the necessary equipment. Scales are essential, so too is a cloth on which to judge comparatively.

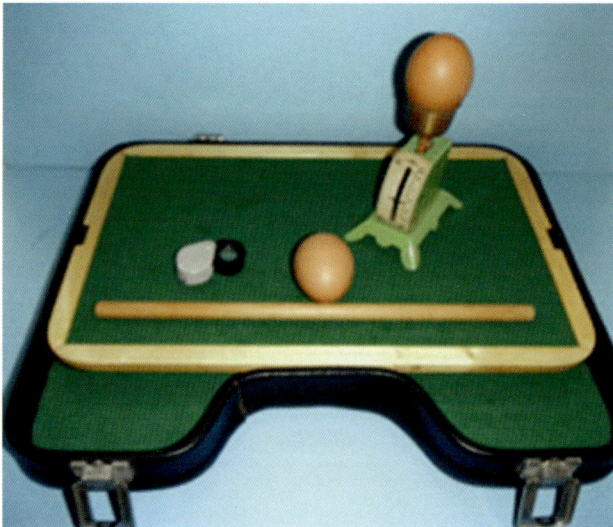

Egg judging kit

Green will serve for all colours although I prefer to use a black cloth for white and cream eggs. A piece of wood to line up the eggs against is useful. For contents you will need a sharp thin bladed knife and a cloth on which to wipe 'eggy' hands. All my judging aids are transported in a purpose made box constructed by my son for a GCSE D.T. project.

Judging Procedure for Externals

Before judging, you and your steward should go through the schedule, checking the classes, disposing of empty plates and checking eggs obviously placed in the wrong classes. In doing this, you must abide by the schedule and as a matter of courtesy, inform the show secretary of what you are doing.

My first judgement is always based on an impression of size and shape. This is taken standing well back to view them in one glance. I then move those half dozen or so plates I believe will be in contention for the awards, to the front of the table and move to the back plates that I do not like immediately.

The next step is to handle the eggs, testing them for freshness and shell texture; failure to satisfy these points means removal from the 'front bench'. The remaining plates are then compared for matching on the judging cloth. As I move the eggs to the cloth I take the opportunity to have another look at the shell texture and to note colour application. If necessary I make use of a X10 magnifier to check blemishes from dirt or my imagination. For white eggs, I use a black cloth, for tinted and brown eggs, a green cloth.Comparing eggs against a uniform background makes it very easy to see bulges and other shape defects. Lining the sets of eggs end on to a round wood dowel helps check their length. The rod can also be placed on top of the set of eggs and gently moved to check they have a similar girth. It is surprising how often one's eye has been deceived, an apparently 'matched' set failing to roll together.

I never have more than three plates on the cloth at any one time, for fear of muddling them up. I start with my possible first prize winners and compare them side by side with the second. Should the order be confirmed the former are returned to their plate and third set brought onto the cloth.

This procedure soon becomes second nature and one can continue, even when interrupted, as one frequently is by interested bystanders. Having judged to the required number of places, I then stand back to survey the front row line up by eye as a check on the sequence.

Before entering the order on the judging slip I have a quick inspection of the rejected plates, just in case there is one worthy of a placing. When giving the steward the order, I will get him to note instances where plates were close as to what my deciding factor was and to 'star' the winning plate number if I consider it worthy of the best in show award.

Finally, if I had cause to pass eggs because they were stale, over-prepared or over 1½oz, if bantam eggs, I would indicate this on a slip of paper left on the offending plates. This is done to help both exhibitor and spectator to understand the raison d'etre of the judging.

Judging Procedure for Contents

Contents classes require a sharp knife and a steady hand. Just tap the egg behind the dome at the widest part to crack the shell. With your thumb and forefinger down the blade so only 4 mm protrude insert the end in the crack and rotate around the egg in a chopping motion. When the circle is almost complete quickly pull the shell apart and allow the contents to fall on the plate. Give the contents time to fully fall, especially waterfowl eggs whose contents are rather viscous. Inspect the position and size of the air sac, place the bottom half of the shell over the top and if the sac is faulty place face down on the plate. If it is OK, place the shells on their side so the sac is still on view for future judging. If the yolk or chalazae appear tangled give them some sharp puffs of breath to straighten things up. When all the eggs are open, I walk away to judge elsewhere to give 10 minutes for the contents to settle. Remember that the 1½ oz rule exists for contents as well as externals – heavier eggs should be left intact. I judge the contents in the same way as the externals, moving the best plates to the front of the table – take care!

Judging the External/Internal Class

Some shows have a class for external internal and provide cards on which to score the points against the standard. This is quite a time consuming business. I have two 'short cuts'. In one I score the features in the ratio of the standard points thus:

Externals – Shell, 5; Size, 3; Colour, 4; Texture, 4; Freshness, 4
Internals – Yolk, 6; Albumen,6; Chalazae, 2; Air Sac, 2; Freshness, 4.

I find this simplifies the addition and provides less scope for worrying, for example about the difference between 25 points or 23 points for yolk. In the other method (and this is only possible in small classes) I arrange the externals in rows or groups according to their quality – best front left and so on. I then open the eggs and move the contents within the rows. An excellent content could move up to be in the cards, a poor yolk could move a well placed egg down. This method does rely on having a good memory!

Judging Best Eggs

At the end of judging you will be required to give specials and choose a best exhibit in show. It is best to be aware of the specials at the start so you can mark them up as you complete the specialist classes. I also 'star' any exhibit in my judging book that I feel might be in contention for best exhibit. If there are a lot of classes ask for a spare table and move the plates that are in contention on to it for a close comparison. Remember contents can win, so can a single egg but they would have to be excellent for their maximum points are 100. For every matched egg there are 5 extra points, a plate of 12 has a possible 160 points. At present there are standards for displayed, painted and decorated eggs but they cannot qualify for best eggs. Bear in mind, also, the breed requirements for the specialist classes. Welsummers must be dark brown, ideally 'flowerpot red', Araucanas an even blue and so on. Finally once your decision is made stay around and check that the cards correspond with your placings – administrative errors do occur. It also provides exhibitors a chance to discuss with you the exhibits on the tables.

Poultry Club Judging Test

After judging a few shows you may then consider entry to the Poultry Club Panel of judges. This involves attending the National Show and taking a two part examination. The theory section, which lasts about 30 minutes, tests your knowledge of the egg and the Poultry Club rules for showing. The questions are all 'tick the box' type. This is then followed by a practical test with an experienced egg judge. You will be asked to judge several classes under his/her eye and then justify your placings. This takes between 40 minutes and 1 hour. To qualify you have to pass both sections and achieve a mark of 70% overall. Once admitted to the Panel you are placed on the official list which is printed in the Poultry Club Yearbook.

Chapter 19

SPECIALIST EGG CLASSES

S ome breeds of chicken have retained an emphasis on 'utility' in their standard. The most obvious examples are the Marans, a French breed, which was introduced into this country as a white flecked table bird in 1929 and the Welsummer, introduced from Holland in 1928. The Marans has 20 points for utility in its standard and the Welsummer 30. Both lay a dark brown egg, although in the case of the Marans this was not a factor in the decision to import the birds to Britain.

The Welsummer

The Welsummer Club is the only breed Club to have an egg standard published in the Book of Standards. The points awarded differ from that

Welsummer flower pot reds

of the General Egg Standard: 25 points are given for colour, size gets 20 points and freshness and bloom 10 points.

When the birds were introduced great emphasis was paid to the egg which was large for the size of the bird, up to 3 ozs, and dark brown. The type of bird that laid these eggs, however, were very variable. Discussions as to the 'correct' standard were quite heated! The colour in some strains is a very deep maroon/purple and very striking they are on the egg table. The more usual type is a deep red/brown, 'the flowerpot reds'. Both these types are usually seen with a matt finish. Some strains provide speckles of a deeper pigment, others blotches. Whatever the colour the egg should be of good size yet retain the required shape. So often large eggs show bulges and uneven symmetry. The ideal should exceed 2½ ozs.

As much of the colour is applied with the cuticle of the egg it is very prone to being dislodged or smeared in the process of laying. These 'scratch marks' are more likely to occur if the nest box material is unsuitable. They can be avoided by using soft wood sawdust and chopped soft Barley straw. Do not use Wheat straw, it is too hard, or hay which attaches to the cuticle and leaves ugly marks when removed. Regular removal of eggs from nest boxes will prevent damage by subsequent layers.

As the colour is so intense problems do arise over consistency. Watch out for uneven application, the most common being a dark top and light end. Eggs at the start of the season are darker than those laid at the end. Indeed, variation can occur within the individual clutches. Prolific layers tend to lose the colour more quickly than the occasional layer.

To protect the colour from fading the eggs should be stored out of the direct light. When transporting the eggs protect them in tissues for egg cartons can ruin a prize exhibit.

Remember that eggs shown in the mottled classes are judged on the depth of ground colour and evenness of mottle. In Welsummers, the prize winners should retain an even depth of dark ground colour.

When breeding for colour select from the hens that consistently lay good eggs and use a cock whose progeny are proven good layers. Most breeders emphasise the importance of the male in maintaining good colour.

The miniature birds initially were developed from game crosses and not surprisingly the egg colour was poor. This is still the case some 60 years later. The only way to maintain the colour is to breed down from good large stock and this takes time. It does help to explain, however, why one sees so many oversized bantam Welsummer eggs on the show tables. To conclude I quote from the Secretary's report to the newly formed Club in 1930:

"It is the rich brown egg which makes the Welsummer distinctive. Future popularity rests on that egg and every breeder should breed and retain it as a characteristic. Lose this asset and what has the breed to offer in competition with other established breeds?"

The Marans

Much that has been said above about the Welsummer can apply to the egg of the Marans. In general, however, the colour is not quite so intense and it veers towards a yellow/brown rather than purple/brown or red/ brown. The eggs are usually glossy and lack the size of the Welsummer. The miniatures produce a very attractive egg, well within the size limits, and of a true brown shade. Being gloss Marans eggs are a little less susceptible to scratch marks than Welsummers – it does, however, depend upon the strain. Harold Britcher, a very successful exhibitor of Marans eggs in the 70's and 80's reflects thus:

"With some strains of Marans the brown colouring is more embedded in the shell than in others. Some, although a

Marans bantam

very good dark brown have the colour only on the surface of the shell and it will be found that this can easily be removed if the egg is rubbed.

It needs to be stressed that the nesting area must be large enough for the layers to move about freely. The maxim should be a large nest and a small number of birds using it. Quarrelling over nesting space is a contributory cause to the eggs' surface becoming marked as the beaks and feet of the jostling birds touch it.

It is also important that the egg does not come into contact with the hard floor of the nest box as the hen settles down to lay. We need a soft cushion of straw between the egg and the nest floor.

Having collected the fruits of our hens' labours it will be necessary for those even slightly dirty to be cleaned straight away. Leaving dirty eggs for any length of time will result in staining and once a Marans egg is stained you can say good-bye to it. To clean an egg one needs to place it in lukewarm water for a minute or so and then, under a slowly running cold tap the egg can be revolved in the hand, lightly touching the offending marks with the thumb. Do not use hot water as this can often damage the colour. Bits of straw stuck to the shell can usually be soaked off without leaving a mark. Now the egg must be carefully dried whih is done by merely patting with a soft towel, never rub the egg or you will probably do damage.

In saving them for show, keep all the likely winners and make the final selection in good time. Eggs need to be as fresh as possible but this is not always practicable as your special hen that is producing the goods may keep you waiting for some days. A cool dark place with the eggs in cartons, soft tissue paper covering them and they are alright for seven days or so. Never let the egg see the light of day till show day or the colour will probably fade."

Fading in Marans is very obvious to an experienced judge for the shiny bloom is lost, the shell taking on a lack lustre green/brown shade. (See page 74) Such a loss of colour is very obvious at the summer Agricultural Shows when exposure to bright light accelerates the process. Rarely can eggs be shown again after a day in a hot tent.

Of the two breeds the Marans must be considered as the more successful on the show bench. Good miniature eggs double the chances of top awards. That said a first class plate of Welsummer eggs does take some beating. Such success, however, is not so often seen in contents classes. In my experience it is the brown eggs that produce the blood and meat spots as well as showing a greenish hue to the albumen. Care does need to be taken over feeding the dark egg layer for contents in order to obtain the right balance between yolk colour and the albumen. Prior inspection of the hens egg contents is essential, so is your ability to recognise that hens egg next time she lays!

Over the last few years French Marans have been imported into this country. The breed is now accepted by the PCGB standards committee. The eggs are striking as they are a very deep, dark brown however their poor shape currently lets them down.

The Araucana

In both the breeds discussed so far the colour has been deposited on the surface of the egg but with the Araucana the blue pigment is seen

Araucana eggs

throughout the shell. As stated in the chapter on the shell this is due to a single gene effect, unlike the brown colour which is the result of several genes interacting. The colour ranges from a greyish/violet blue to a turquoise or greenish blue. Sometimes eggs are seen that are green rather than blue or even olive. At the present time there is much debate in the Araucana Club as to whether these colours should be accepted. The argument being that they indicate crossbreeding. It is well documented that crossing with brown egg layers produces green, olive and khaki egg colours.

The Araucana is not a big breed so not surprisingly the eggs of the large fowl are small in comparison with Buff Rocks and R.I.R.'s. Their shape, however, is generally quite acceptable, more so than the miniatures. Their eggs are often up to the size limit and suffer from translucent patches in the shell. When one does see a good plate of Araucana eggs they are very eye catching. Many egg exhibitors keep a few Araucanas to provide the third colour in the plate of three distinct.

The Croad

These then are the three main specialist breed classes seen at most shows. At large shows, such as the National, other breeds have egg classes. Literature from last century reports on the dark eggs of the Asiatic breeds but now only the Croad Langshan upholds this feature. Breeding for other features has lost the egg colour. Croads at their best demonstrate the 'plum' coloured egg, a mauve brown, but this is rarely seen today.

The Barnevelder

The other breed in which breeding for feather has caused the loss of egg colour is the Barnevelder. A Dutch breed like the Welsummer it was introduced into this country in 1921 for its dark brown eggs. Some breeders are trying to re-establish the colour of the egg which at present in most cases is really 'dark tinted'. However the tint is clearly brown.

These then are the five specialist classes. It has to be said that other breeds can lay eggs that will challenge for best in show – white eggs from the Ancona, Hamburgh and Leghorns or light brown eggs from the Australorp, Buff Rock, Rhode Island Red and Sussex. All take their share of the prizes.

Two examples of decorated eggs

Elaborate decorated egg   **Child's decorated egg**

Chapter 20

DECORATED, PAINTED AND DISPLAYED EGGS

The summer Agricultural Shows usually provide egg classes for those who have an artistic eye. Variously classified as painted, decorated and displayed. They provide an attractive display for the public to view. After the public been surprised by birds with feathered legs, with heads like pom-pom dahlias and combs like squashed strawberries, these egg classes provide exhibits that fall more within their experience. Often supported by the W.I. and schools, the classes are very competitive. At the Royal Cornwall Show I was once confronted with 138 exhibits, 98 in one class alone!

How does one judge these classes? There is no recognised Poultry Club standard to help and I do not know of any rules laid down by any other body.

If it is a large class with eggs variously produced, I group them into the categories mentioned above and award separate prizes. The groups are, I feel, quite distinct and hopefully show organisers reading this article will adjust their future schedules to accommodate the separate classes. My criteria for the classes are as follows:

Decorated Eggs

Here the egg is often painted, but in addition there is some form of decoration. They range from the very elaborate eggs, whose shells have been cut, revealing painted interiors behind open doors or drawers. In all cases the egg must be seen and it's shape recognisable. Decoration takes the form of beads, cardboard, shells and, of course, colouring. (Left two examples of decorated eggs)

Two examples of painted eggs

Painted Eggs

These eggs, either blown or hard – boiled, are painted with any of the usual media, oils, watercolours, inks etc. There is no other adornment whatsoever, not even beads for eyes or bits of extra shell to provide texture. To set them off nicely, the organisers should provide rings about ½" deep in which the egg can be placed upright.

Group of four painted eggs

Displayed Eggs

Unlike the other two classes, this is usually for several eggs, usually six, the eggs themselves being unadorned. The exhibit consists of suitably chosen materials, cloth, flowers, bark or cardboard that serves to highlight the clutch of eggs within the arrangement. The pictures show well how natural material can be used in this way.

It is only in the displayed eggs that I give any credit to the eggs used as, ill matched, poor textured eggs would not enhance the overall effect of the exhibit. I would not, however, penalise an arrangement that included bantam, fowl and duck eggs, providing their shape was a match and their shells an even texture. Different colours, too, would not be penalised if such contrast were helpful to the general effect.

Two examples of displayed eggs with the eggs clearly visible

Showing these eggs is fun, providing a welcome relief from the more serious show classes. Judging is subjective and cannot be placed under the close scrutiny that some of the other judging decisions are by those convinced their exhibit is best! It is the judges' opinion that makes the award on the day, even with the scale of points that I have outlined opposite.

In the Champion Egg Shows these classes compete for their own Bronze Star Award card.

I hope this brief account might encourage a few more to take up the art of painting and decorating eggs during the long winter evenings and so provide a little more colour and interest to the egg displays at the summer shows.

In 2007 the poultry Club accepted a standard for these classes. They followed the scale of points suggested opposite. The standard also accepted that points for egg quality would only apply to Displayed Classes.

Suggested Judging Points

Painted Eggs:

Impression/artistic effect 25 pts

Originality/concept/
    subject matter   25

Quality of painting   25

Use of colour   25

            100 pts

Decorated Eggs:

Impression/artistic effect 20 pts

Originality/concept/
    subject matter   20

Quality of construction   20

Use of colour   20

Use of materials   20

            100 pts

Displayed Eggs:

Impression/artistic effect 30 pts

Originality/concept/
    subject matter   25

Use of materials   25

Quality and matching of
    eggs used   20

            100 pts

The judge will assess egg quality, shape, colour and matching

Chapter 21

WATERFOWL EGGS

When entering up the results of the 1988 championship shows for the Poultry Club Year Book I was interested to note that six best egg awards had been won by duck eggs. This compared with seven wins by hen eggs, four by bantam eggs and two by plates of six eggs. To some egg exhibitors this should not be, like poultry men who object to waterfowl winning best in show, they feel comparisons of such unlikes to be invalid.

Prior to the 1997 revision there was no mention of waterfowl eggs in the standard! Without a standard waterfowl eggs could not qualify.

However, waterfowl eggs did get a mention in the Poultry Club rules. I quote:

"(c) Separate classes must be given to waterfowl, turkeys and guinea fowl.

In contents classes waterfowl eggs should be exhibited in a class for waterfowl and not with large fowl or bantam classes."

These rules imply that waterfowl eggs are different from those of chickens which provides further evidence for those opposed to 'mixed' egg showing.

So are waterfowl eggs different? Obviously they differ in size. Hen eggs weigh between 1¾ and 2¼ ozs (average 52 g). Bantam eggs have an upper limit on the show bench of 1½ ozs (42.5 g). Romanof states in his book *The Avian Egg* that the average weight for a duck egg is 2¼-2½ ozs (67 g) and that for a goose is 6¼ ozs (177 g). Bantam duck eggs that I have weighed indicate an average weight of between 1½ and 1¾ ozs (48 g).

These increased weights are not wholly due to the larger size of waterfowl eggs. They have thicker shells and denser contents.

**Table 13 Comparison of the Percentages of Water
and Solids in Poultry Eggs**

	Chicken Egg	Duck Egg	Goose Egg
% Water	73.6	69.7	70.6
% Solids	26.4	30.3	29.4

The 3-4% reduction in water content makes the contents noticeably more viscous. The solids are directly related to the potential energy in the egg. So it follows that per 100 g the duck egg is the most 'nutritious' with about 200 K calories. This difference in calorific values is no accident. Waterfowl eggs, especially ducks, are laid in a nest very near to water so they would be damp and cold and the embryo would have greater heat loss so it would have to 'burn' more food reserves. Another relevant factor may be that waterfowl have a slightly higher body temperature than hens.

Duck Eggs   **Goose Eggs**

Shape is of paramount importance when judging eggs, the ideal is broad at the dome narrowing equally on the sides to a rounded end. In comparison with this, the quoted average length and breadth of goose eggs indicate that they are proportionately longer and so appear narrower and more pointed than the ideal. The same is true for the Mallard, and so, I suspect, bantam ducks would be naturally longer and thinner too. This is borne out by my observations when judging; rarely have I seen either a goose or bantam duck egg that scored well for shape. If this is so, it seems unfair to judge these waterfowl types to the poultry standard. Duck eggs, however, have average measurements similar to the hen ideal so, not surprisingly, it is they that won the awards. Runner duck egg proportions match the best, followed closely by eggs from Pekin ducks and Muscovies.

Even application of colour is easier to obtain in waterfowl eggs than say the dark browns of the Marans and Welsummers as most types lay a white egg; either a brilliant white, as in Pekins; or a greeny white, as in some Aylesburys. The exceptions are the Rouen and the Rouen Clair which lay pale green eggs, the Blue Swedish which lay pale blue eggs and the Cayuga whose egg is a very dark green, in some cases almost black.

Bantam ducks will produce eggs of varying shades of green/white depending on how close they are to the ancestral Mallard. Just as nest marks spoil the colour on chicken eggs so water stains spoil the colour on waterfowl exhibits.

For many judges it is the shell texture that makes a duck egg stand out. Their smooth texture and purity of colour catch the eye. Though it must be said that some ducks produce a 'colour wash' effect within the shell and some show rather unsightly air sacs. Muscovy eggs are laid covered with a waxy coating which scratches easily. If this is removed by 'scrubbing' they then exhibit a very fine shell texture. Here some would argue that this constitutes 'over preparation' as defined by Poultry Club show rules. In contrast to duck eggs those of geese are often coarse in texture and rather 'chalky'.

The foregoing differences in the external features of waterfowl and chicken eggs may escape the untrained eye but the differences in their contents will surely not remain unnoticed. Their albumen is altogether more viscous, the well defined thick white rarely failing to support the yolk high and centrally. The clear albumen shows the chalazae to good effect and is a marked contrast to the firm yellow yolk. If competing against chicken eggs waterfowl would win nine times out of ten!

The increased popularity of waterfowl has prompted many shows to stage waterfowl egg classes and, as I stated at the beginning of this chapter, the eggs are showing well. They are, however, different from chicken eggs in several important respects.

Waterfowl contents
Note the clear and viscous thick albuben

Chapter 22

TURKEY EGGS

The formation of a Turkey Breed club in 2000 raised the profile of turkeys and over the last fifteen years they have reclaimed a place in the poultry fancy. Most shows now stage turkey classes. More recently there has been a demand for separate turkey egg classes at shows. In 2014 the poultry club recognised this by providing Bronze awards for best plate and best individual turkey egg.

The current poultry club standard book covers all eggs within one standard and all have the same scale of points. It does, however, recognise turkeys and waterfowl as distinct and requiring of separate classes.

Turkey Eggs

How do turkey eggs differ from chicken eggs

As turkeys are larger than chickens they lay a larger egg. The wild turkey has an egg weight of 75 g compared to 85 g in selectively bred white Holland turkeys. Marsden & Martin (1955) provided a detailed analysis of size suggesting seven categories from very large – 106 g to very small 64 g. Hatching data from 8000 bronze turkey eggs indicated that hatchability was highest in eggs weighing 71 g to 98 g (2½-3½). In a study of 4341 light breed eggs the heritability was best in eggs between 73 g-87 g (2½ oz-3 oz) This suggests that the standard weight for turkey eggs should be 70 g to 98 g (2½-3½).

In addition to the weight difference between turkey and chicken eggs there is also a difference in the weight of the parts within the eggs.

The Average Weight of the Parts of Turkey Eggs
(Asmundson, 1939)

		Turkeys	Chickens
Part of Egg	*Weight in Grams*	*Weight of the Part as a Percentge of the Egg Weight*	*Weight of the Part as a Percentage of the Egg Weight*
Yolk	27.68	30.53	29.91
Albumen	54.17	59.74	59.73
Shell Membrane	1.32	1.46	0.96
Shell	7.50	8.27	9.40
Total	90.67	100.00	100.00

The differences in the percentage weights of the yolk and albumen are hardly significant. The differences in percentage weights of the shell and membranes are more interesting. Anyone involved in hatching turkey eggs has experienced the tough nature of the membranes. There is nothing in the literature to suggest that the composition of the shell and membranes differs in turkeys and chickens. Could there be a genetic reason? Turkeys incubate larger clutches than chicken – 15 to 20 as opposed to 8 to 12. Fertility drops off as eggs get older, in chickens it is not recommended to hatch from eggs more than 10 days old. In turkeys the oldest egg will be older than this, so are the differences related to retaining egg 'quality'? Does the increase in membrane density protect from entry of bacteria or even slow evaporation?

The current standard states that turkey eggs tend to be more conical than chicken eggs. While this may be so of eggs seen at shows I am not convinced that they are representative of the general turkey population. Viewing pictures on the world-wide web indicate a variable egg shape. As the 'Cassinian oval' with an index of 74 is the ideal for hatching chicken eggs there is no reason why this should not hold true for turkey eggs which hatch under similar conditions - unlike waterfowl eggs.

Showing Turkey eggs

The criteria for exhibiting in turkey egg classes are the same as the other types of egg. The eggs need to be of the standard shape, size, and shell structure. A strong shell free from calcareous blemishes is especially important bearing in mind the longer incubation period of turkeys.

The colour of the turkey egg sets it apart from other breeds as they are always speckled; the specks are reddish/brown on a ground colour that ranges from off white to a deep cream which sometimes has a pinkish tinge. This makes for an attractive egg. For show purposes the more uniformity in colour and density of speckling the better. Blotches or mottling of colour could be considered a fault. As with all classes of specked and mottled eggs the evenness of ground colour is important. If no turkey classes were scheduled they would be shown in the class according to ground colour – cream.

Freshness in turkey eggs is more difficult to assess by eye as they do not have the obvious bloom shown by some breeds nor an overall colour that can fade. A judge is empowered to break one egg in a plate to check for freshness. I prefer to have a glass of water to hand and see if the egg sinks.

As in all plates of three or six there are 5 extra points awarded for each egg that matches. In my experience individuals lay a consistent shape which makes selecting eggs an easier task. Only as the laying season progresses do the intensity and distribution of speckles vary. A well matched plate of turkey eggs does catch the eye and such exhibits are now regularly in consideration for the 'Best Egg Exhibit'.

Internal Structure

Apart from the tough membrane already referred to the internal structure and chemical composition of the turkey egg is very similar to that of chickens. The differentiation into thick and thin albumen varies with each individual and the colour of the yolk is influenced by the con-stituents of the diet. When the contents are inspected the germinal spot (blastodisc) shows as a chalky white spot which should be uppermost due to the action of the chalazae. In a fertilised egg the germinal spot develops into the blastoderm which is characterised by a transparent halo. This is very noticeable in chickens but less so in turkeys making decisions on fertility more difficult

At the time of writing the Turkey Club are considering a Turkey Egg standard for acceptance by the Poultry Club.

SECTION 4
EGG MISCELLANY

Chapter 22

EGG FACTS

Size

The Largest Eggs in the world are laid by the ostrich. They measure some 15-20 cm. in length, are 10-15 cm. in diameter and weigh 1.65 to 1.78 kg. (this is equal to about 2 dozen hen's eggs). They require a boiling time of 40 minutes. The shell, 1.5 mm. thick can support a human weighing 127 kg. The largest eggs laid by British birds are those of the mute swan. They measure 109-124 mm. in length and 71-78.5 mm. in diameter. The weight is 340-368 g. (an average hen's egg of about 58 g.).

The largest egg on record is from the Elephant bird (Aepyornis maximus), also known as the 'Roc bird', which lives in southern Madagascar. It was a flightless bird.

One huge egg example preserved in the British Museum (Natural History) measures 856 mm, 33 in. round the long axis with a circumference of 723 mm, 28 in., giving a capacity of 8.88 litre, 2.35 gal, or seven times that of an ostrich egg.

The largest egg laid by any bird in proportion to its own size is that of the Kiwi (Apterygiformes) of New Zealand. There is an old record (1889) of a 1.68 kg 3 lb 12 oz Mantell's kiwi (Apteryx australis mantelli) laying an egg weighing 406 g 14.5 oz, or nearly one-quarter the body-weight of the hen.

The Smallest Egg recorded came from an Helena's hummingbird found in Cuba. It was 11.4 mm. in length, 8 mm. in diameter and weighed 0.5 g. The smallest laid by a British bird are those of the Goldcrest, 12.2-14.5 mm. long and 9.4-9.9 mm. in diameter.

The smallest egg laid by any bird in relation to body-weight is that of the Emperor penguin (aptenodytes forsteri), which constitutes only 1.4 per cent of the total body-weight

Heaviest

A Hen's egg of 454 g. with a double yolk and double shell laid by a White Leghorn in New Jersey, U.S.A. in 1956.

The heaviest hens egg reported in the UK was one of nearly 339 g (12 oz) for an egg with five yolks. It measured 31 cm (12½ inches) round the long axis and 22.8 cm (9 inches) round the shorter axis. It was laid by a Black Minorca at Mr Stafford's Farm, in Mellor, Lancashire, in 1896.

Numbers of Eggs Laid

In the World: 361 in 364 days by a Black Orpington in New Zealand in 1930.

In the U.K.: 353 in 365 days by a Rhode Island Red at Milford in Surrey in 1957.

For a Flock: An annual average of 313 eggs per bird in a flock of 1,000 at a farm in Albury, Surrey, in 1957.

Other Egg Records

Most Yolks: 9. From two hens, one in the U.S.A., one in the U.S.S.R.

Largest Easter Egg: 3.11 m. high 7.54 m. in circumference, weighing 2,035 tonnes, made in Melbourne, Australia.

Egg Dropping: The greatest height from which fresh eggs have been dropped without breaking is 182 m., this from a helicopter.

Egg Shelling: 1,050 dozen in 7½ hours by two blind kitchen hands in Trowbridge.

Egg Throwing: Without breaking, 316' 5" in Missouri, U.S.A.

Largest Omelette: Reportedly made with 11,145 eggs, taking 21 chefs 1 hour to make, in Bradford, West Yorkshire.

Omelette Making: 188 2-egg omelettes in 30 minutes by a chef from Michigan, U.S.A. He used 6 pans and 6 burners.

Gluttony! 14 hard-boiled eggs eaten in 58 seconds, in Corby. 31 soft-boiled eggs in 78 seconds, in Northampton. 13 raw eggs in 2.2 seconds, in Norwich.

Most first prizes: In 1999 George Taylor from Cumbria gained 536 red cards from 82 Poultry Shows in year.

Chapter 23

EGG FABLES AND SAYINGS

Well known sayings
- Teach your grandmother to suck eggs
- Sure as eggs are eggs
- Don't put all your eggs in one basket
- Better half an egg than an empty shell
- Cackle often but never lay an egg
- You can't unscramble an egg
- Nest egg – give a child an egg. The chicken that hatches from that egg is sold and the proceeds put aside for the child.

Eggs and religion
- The 'Mundane' Egg – the egg from which the world was hatched (Egyptian, Hindu, Japanese)
- Anglo-Saxon goddess of spring Eoestre (Easter) holds an egg – the symbol of life.
- Chinese dye eggs to celebrate Spring and give red eggs to children which symbolises long life and happiness.
- Christian – during Lent forbid eating of eggs, on Good Friday eggs blessed in church before being eaten to symbolise new life also at Easter eggs were hard roasted for children to play with.
- 'Pace eggs' (Passover). Eggs also hidden in trees and bushes for children to find.

Egg Fables
- In the middle ages people believed witches (who changed size) travelled in empty eggshells. They therefore crushed them.

- In Cornwall a belief that egg albumen on the skin will cause warts.
- An egg laid on Ascension Day if hung in the roof will protect the household.
- Round eggs produce cocks: eggs laid on Mondays and Fridays produce hens.

Eggs cures for ills
- White of egg for headaches
- Boiling eggs in donkey urine to cure kidney disease
- Sucking eggs for coughs
- Raw egg on beet leaf to soothe burns

Egg Art Forms
- Faberge eggs – fabulous jewelled eggs just prior to Russian revolution
- Nuremberg eggs (1700) gold yolk, enamel chick with jewelled egg within which is a ring – classic 'surprise' egg

Chapter 24

ABNORMAL EGGS

Double eggs

These result from faulty movement down the oviduct. The development of one egg becomes mixed up with the development of another. Various types exist.

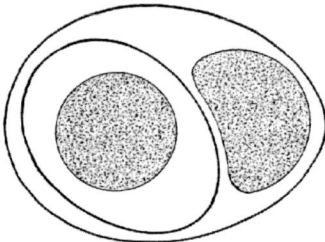

Egg within Egg

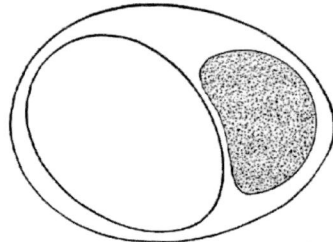

Yolkless egg within egg

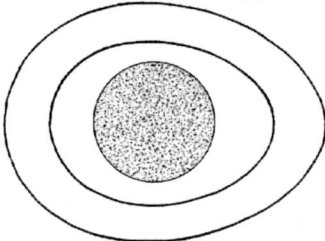

Egg within yolkless egg

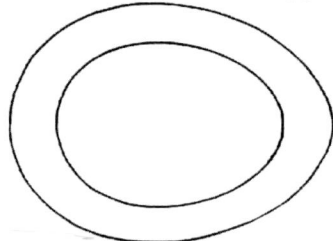

Yolkless egg within yolkless egg

Fig 27 Common forms of Double Eggs

Double Yolked Eggs

These are comparatively common. Two yolks that are the result of double ovulation move into the oviduct together and are processed within the same albumen and shell

Dwarf Eggs

These are also known as witch eggs, cock eggs and wind eggs.

They are very small and usually do not contain a yolk. They are often laid by pullets before starting to lay properly.

Some form because the ovum that has ruptured from the ovary fails to enter oviduct. The oviduct, however, continues to secrete albumen and a shell.

Freak Eggs

These are of varied, often bizarre shapes. There is no common reason for their occurrence. In those that are clearly misshapen eggs a malfunction or malpresentation within the shell gland must be involved. Some egg shows have classes for *'the most unusual shaped egg'*.

Examples of misshapen eggs

Other misshapen eggs obviously have extra pieces of membrane or shell. The shells are often brittle. The eggs usually have contents.

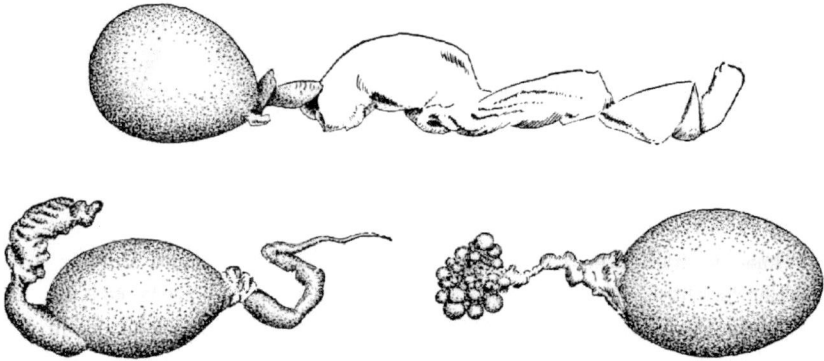

Fig 29 Eggs with various forms of external attachments

Soft Shelled Eggs

Such eggs are quite common and are due to a reduced shell deposition. Some have hardly any shell and are very pliable, others have a thin shell coating and are brittle to touch. In a few cases the fault is genetic but most are laid at the start or end of the laying season. Older birds are more prone to produce soft eggs. There is either a lack of calcium salts with which to make the shell or the bird has encountered stress which has upset the laying rhythm. Oyster shell was traditionally given to birds to prevent soft shells.

Truncated Eggs

These eggs have normal contents but their final shape is 'slab sided'. This is the result of abnormal pressures in the shell gland during the early stages of shell development. Most of the flattening is towards the pointed end of the egg. One suggestion is that the leading pointed end of the affected egg comes up against a previous egg that has been held back in the shell gland.

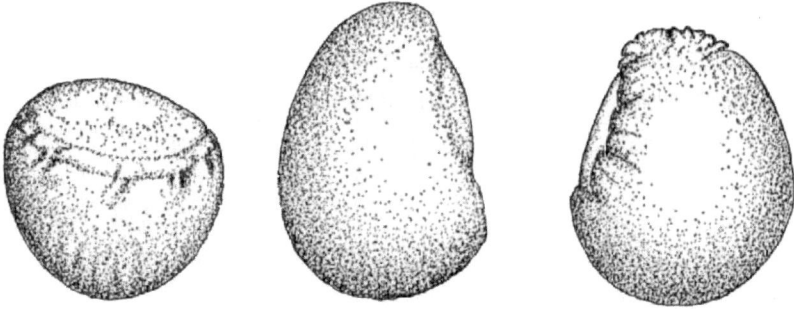

Fig 30 Examples of Truncated Eggs

RECENT ADVANCES IN AVIAN EGG SCIENCE

Philippe B. Wilson MChem(Hons) AMRSC

As breeders and judges, it is important to remain constantly apprised of novel approaches, techniques and discoveries relating to poultry and their husbandry. The inclusion of this chapter in the second edition is to encourage readers to regularly search for updates and research on areas that may impact their breeding or showing strategies.

It is well-known that eggs form an integral part of agronomic economy and our own feedstuffs. It is therefore not unexpected to find that research into egg structure, production and colour is constantly being undertaken.

One area of key interest is egg shell colour and structure. It has been proven by numerous studies that the colour of eggs significantly affects their popularity in commercial terms. It is therefore surprising that eggshell pigmentation of wild species is a far more treated subject that that of the brown egg-layer.

Brown Egg-Layers: A Case Study

Originally, the brown pigment extracted from eggshells by Sorby in 1875, was called oorhodeine, however it was not before 1973 that Kennedy and Vevers concluded that protoporphyrin IX was in fact responsible. Protoporphyrin IX, biliverdin, coproporphyrin and uroporphyrin are all found in the shells of domestic egg-layers, accounting for the complexity in the colouring of egg shells we see from domestic breeding. In fact, it has recently been confirmed that egg-shell colour has significant heritability and links have been found between brown-pigmented egg shells, and egg strength, degree of hatchability, and bacterial resistance.

Scientifically, the colour of eggs is measured using a reflectivity meter. This device detects the degree to which light is reflected from the egg shell and therefore detects the darkness of shell colour, with a dark shell having a low value and light shell, high. Although it begins to quantify shell colour, Shell reflectivity is not the most accurate method of detecting the colour of eggshells. This is often suggested as being due to the small sampling of the method, which only measures a small area on the egg shell.

For more quantitatively accurate results, spectrophotometry is used instead of reflectivity. This measures a value between 0 and 100, where 0 corresponds to black, and 100, white. One advantage of spectrophotometry is that a much larger area of the shell is considered, which by nature will give overall more appropriate results.

There are still issues arising with measurements of shell colour. Some shells are covered in differing amounts of calcium deposits, which shield the pigment from the sensor.

In fowl, the protoporphyrin IX pigment is mostly found in the outer skin cells in the shell gland. Hormones have been found to significantly contribute towards the production of pigment, among them oestrogen, progesterone and prostaglandins, although work is ongoing to detect the exact location of pigment synthesis in domestic egg-layers. Studies have been carried out, which have shown, for example, that pigments are made in different sites in Rhode Island Reds, compared to White Leghorns. Porphyrin pigments were present in Rhode Island Red uterine tissue, but absent in that of White Leghorns.

It is known that every bird within a species will secrete and deposit pigment at difference times, and that these will be different in blue, brown, and white egg-layers. Japanese quail are a species which have undergone more research than commercial hens, but have similar uterine structures. Pigment is deposited on quail eggs approximately 3 hours before laying, whereas some blue egg laying fowl have been found to have pigment deposition occur at approximately 2 hours before lay.

Warren *et al.* assumed that pigment for brown shelled eggs is deposited throughout the formation process, yet 50 to 74% has been discovered as being deposited in the last 5 hours before lay. Hormones and other factors such as the mineral content in the shell gland have been stated as responsible factors in the finishing of the cuticle layer and the termination process. These also affect the deposition process of the pigment, around which research is still ongoing.

The relationship between Nutrition and Egg Colour

Improving the colour of eggs has long been of commercial and exhibition interest. With consumers now searching for quality over quantity, following the premise that free range is a natural production method, it is interesting to note that it can be more difficult to maintain egg colour in free range flocks than caged.

It is unsurprising that nutrition is a contributing factor to the quality and colour of the egg, both internally and externally. Hooge *et al.* found that feeding probiotics to layers could improve eggshell colour, particularly in brown shelled eggs. Supplements containing *Bacillus subtilis* were administered to 63-week old Lohman Brown commercial hybrids, resulting in improved intensity of the brown colouration for up to two weeks after first delivery. It is not yet clear how this affects the intensity of the colour – some publications postulate the relationship between certain amino acids in the probiotic, with mediation of metal insertion into the polyporphyrin.

It has been noted that iron soy proteinate supplement significantly improves eggshell colour in brown egg-layers, while Vanadium adversely affects pigmentation of the shell.

Vanadium is a first-row transition metal often found in small quantities in poultry feeds. Odabasi and coworkers found that by administering proportional quantities of vitamin C, the adverse effects of the vanadium with respect to shell colour were neutralised. There have previously been suggestions that vitamin D could be responsible for paleness of colouration in shells, however a recent study by Roberts indicated that there was no significant correlation between the two factors.

The chemistry underlying the deposition process is still not completely quantified, however work is currently underway to develop our understanding and provide additional answers to these questions. Genetic factors are therefore difficult to explicitly deduce until such a time, however the first steps towards understanding these have already been taken.

In 1993, Van brummelen and Bissbort published on the control of eggshell colour by several genes which code for various proteins and enzymes regulating the production and deposition of pigments. The specific brown-egg gene, however, remains elusive, although the higher activity of certain key enzymes in brown egg-layers suggests that the brown egg trait is purely of genetic origin.

Wardecka *et al.* established that the same region on chromosomes 2, 4, 5, 6, and 11 influence egg shell colour. Phenotypic observations showed that crosses of white with brown egg-layers resulted in intermediate colours – indicating a codominance effect. The same study found more pigment in crosses of males from brown egg-laying breeds with females from white egg-laying breeds, than the inverse, suggesting a degree of sex-linkage associated with this pigmentation.

The expression of two genes, (SLCO1A2 and SLCO1C1) in the shell gland was found to be associated with the colouring of the eggs. It was suggested by Dunn and co-workers that it may in fact be more productive in the short term to explore the genetic basis of pigmentation by means of selection and breeding rather than molecular methods, until such a time as these become less practically and computationally expensive.

The Effects of Age, Stress and Medication on Laying and Pigmentation

Uniform, dark-brown eggs are the goal of many commercial breeders. It has however been noted that there is a significant, quantifiable shell colour difference between breeds laying brown eggs, and between individuals within those breeds. Although seemingly evident, the fact that blue and brown pigment concentrations are higher within the glands of the respective egg-layers is an indication of the pigments being breed specific, as opposed to being unilaterally produced in all breeds.

Age

Odabasi *et al.* considered a flock of commercial hybrids at different ages, noting the degree of pigmentation of the shell. They found that colour remained constant with the bird, most examples having darker eggs at the start of the laying period, and correspondingly light eggs at the end.

There has been a general trend of decreased pigmentation with age, which has often been observed. A number of hypotheses suggest the increase of egg size with age results in a dilution of the pigment at any point, in order to cover the entirety of the egg.

In 2014, Sammiulah and coworkers carried out a longitudinal study on a flock of commercial hybrids. Longitudinal studies differ from horizontal studies in that factors from the same flock are considered at different ages, as opposed to different flocks at different ages. They found no significant difference in eggshell colour between 35 and 75 weeks, noting however that the colour at 25 weeks was quantifiably darker than other age groups. Regardless of the method of study, the colour of shells became lighter as age increased; the amount of proto-porphyrin IX in 1 gram of whole eggshell from 33, 50, and 67 week old Hyline Browns did not differ significantly. Conversely, the concentration of the pigment in the cuticle layer itself was found to be significantly higher in 50 week birds than those of 33 and 67 weeks. The study there-fore concluded that past optimum laying periods, eggshell colour became paler and that thus age was proportional to paleness of shell in brown egg-layers.

Stress

Reynard and Savory considered the effect of stress from such stimuli as handling and relocation, on the laying trends of brown egg-layers. For 4½ hours prior to oviposition, stress was found to cause a delay in laying of up to 3 hours. A stress threshold was observed whereby hens became unable to lay if the levels and duration of the stress reached or surpassed this.

Moulting stress is an example of a condition or transition severely affecting eggshell colour once laying resumes. Aygun and coworkers showed that the response to moulting varied significantly between individuals in a flock. For caged flocks, it is unsurprising to note that high cage densities are proportional to the level of stress exhibited by the birds, with cage design, fear, and frequency of disturbance also being contributing factors to changes in egg colour.

Mills noted that abnormalities such as paleness, calciferous deposits and some structural defects were related to environmental influences. This was explained by the retention of the egg in the shell gland, where additional calcium deposition onto the formed egg results in screening of the ground colour. Physical stress such as feather removal and ex-tremes of temperature were also confirmed to adversely affect pigment deposition on the shell.

Medication

Drugs can also have a profound effect on both laying and the colour of the shell. Soh and Koga tested the effects of intrauterine injection of prostaglandin F2α on pigmentation of brown eggs laid in the time following administration. They discovered that the hormone accelerated the laying, but produced paler eggs – likely due to the time reduction of the egg itself in the shell gland. They also administered a drug – indomethacin – independently, finding that it entirely inhibited pigment secretion.

Nicarbazin, an anti coddiciosis drug has been noted to result in production of depigmented eggs for up to a week after first dose. The drug concentration and duration of the treatment affected pigmentation, but not the synthesis of the pigments themselves. The effect was found to reverse within a week of the completion of treatment, the birds returning to producing eggs of the same colouration as previously. It is therefore reasonable to conclude that although the pigment is produced under treatment, it is the drug which affects the deposition onto the shell and cuticle, as opposed to its synthesis.

EGGS ON CIGARETTE CARDS

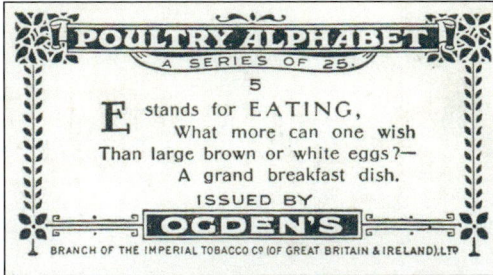

POULTRY ALPHABET
A SERIES OF 25.

5

E stands for EATING,
What more can one wish
Than large brown or white eggs?—
A grand breakfast dish.

ISSUED BY
OGDEN'S
BRANCH OF THE IMPERIAL TOBACCO C? (OF GREAT BRITAIN & IRELAND), LT?

The Poultry Alphabet Set – Ogdens

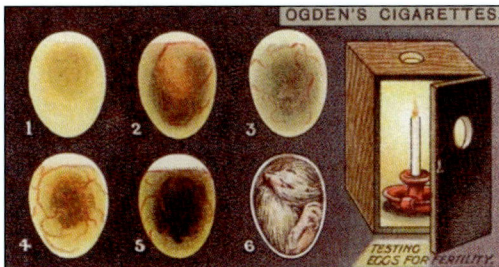

OGDEN'S CIGARETTES.

Testing Eggs for Fertility – Candling

Poultry Rearing and Management Set – Ogdens

How to make an egg tray

BIBLIOGRAPHY

British Poultry Standards 1960

British Poultry Standards, 5th edition 1997

Dr A F Anderson Brown, The Inclubation Book 1979

Robert Burton, The Egg 1987

D A Ede, Bird Structure 1964

HMSO, The Testing of Eggs for Quality 1949

HMSO, Incubation and Hatchery Practice Bulletin 148 1977

J A Oluyemi & F A Roberts, Poultry Production in Warm and Wet
    Climates 1979

Rice and Botsford, Practical Poultry Management 1925

Romanoff and Romanoff, The Avian Egg 1949

S E Solomon, Egg Shell Quality 1990

Lewis Stevens, Poultry Genetics 1995

M G V Thompson, 50 Years of Welsummers 1980

G L Wood, The Guinness Book of Animal Facts and Feats 1976

British Poultry Science Vols. 28, 44, 46

Poultry Science Vols. 21, 40, 59, 70, 80, 84, 85, 92, 94.

INDEX